Mathematical Engineering

Series editors

Jörg Schröder, Essen, Germany
Bernhard Weigand, Stuttgart, Germany

Today, the development of high-tech systems is unthinkable without mathematical modeling and analysis of system behavior. As such, many fields in the modern engineering sciences (e.g. control engineering, communications engineering, mechanical engineering, and robotics) call for sophisticated mathematical methods in order to solve the tasks at hand.

The series Mathematical Engineering presents new or heretofore little-known methods to support engineers in finding suitable answers to their questions, presenting those methods in such manner as to make them ideally comprehensible and applicable in practice.

Therefore, the primary focus is—without neglecting mathematical accuracy—on comprehensibility and real-world applicability.

To submit a proposal or request further information, please use the PDF Proposal Form or contact directly: *Dr. Jan-Philip Schmidt, Publishing Editor (jan-philip. schmidt@springer.com)*.

More information about this series at http://www.springer.com/series/8445

Iman Askerzade · Ali Bozbey
Mehmet Cantürk

Modern Aspects
of Josephson Dynamics
and Superconductivity
Electronics

 Springer

Iman Askerzade
Department of Computer Engineering
Ankara University
Ankara
Turkey

and

Center of Excellence of Superconductivity
   Research of Turkey
Ankara University
Ankara
Turkey

and

Institute of Physics Azerbaijan National
   Academy of Science
Baku
Azerbaijan

Ali Bozbey
Department of Electrical and Electronics
   Engineering
TOBB Economical and Technological
   University
Ankara
Turkey

Mehmet Cantürk
Center of Excellence of Superconductivity
   Research of Turkey
Ankara University
Ankara
Turkey

ISSN 2192-4732   ISSN 2192-4740   (electronic)
Mathematical Engineering
ISBN 978-3-319-83941-7   ISBN 978-3-319-48433-4   (eBook)
DOI 10.1007/978-3-319-48433-4

Printed on acid-free paper

This Springer imprint is published by Springer Nature
The registered company is Springer International Publishing AG
The registered company address is: Gewerbestrasse 11, 6330 Cham, Switzerland

Preface

The Josephson effect was theoretically predicted by Brian Josephson in 1962. Experimental observation of the stationary Josephson effect was firstly realized by Anderson and Rowell, and the nonstationary Josephson effect was realized by Yanson et al. Up to now, there is a growing interest in the fundamental physics and applications of Josephson effect. The developments in Josephson junction (JJ) fabrication technology have made it possible to implement a variety of devices for detecting ultralow magnetic and electromagnetic fields. In addition, discovery of Josephson effect has also enabled the fabrication, testing, and application of ultra high frequency Rapid Single Flux Quantum (RSFQ) based logic circuits for signal processing and general purpose computing.

The Josephson effect still remains one of the most noble manifestations of quantum effects in all of experimental science. In principle, the Josephson effect is nothing more than the electronic analogue of interference phenomena in optical physics. Physical phenomena and electronics applications which are based on Josephson junction still remain challenging at the heart of physics research.

The Josephson effect may be observed in a various structures. The realization of such structures may be achieved in two ways: i) by just fabricating a "weak" inclusion that interrupts the Josephson current flow in a superconductor ii) suppressing the ability of a superconductor to carry a current using different methods such as by deposition of a normal metal on its top, by implantation of impurities within a restricted volume, or by changing the geometry of a sample.

Chapter 1 is devoted to description of theoretical foundation of Josephson dynamics. The emphasis of this chapter is on the general nature of the Josephson effect and on fundamental physical mechanisms that control the current-phase relation (CPR). At the same time, some details are provided regarding types of Josephson junctions (JJs) and their fabrication. We present briefly derivation of Josephson relation using tunnel Hamiltonian method, Ginzburg–Landau (GL) equations and Bogoliubov de Gennes equations. Second section contains results of investigations the influence of anisotropy and multiband effects of superconducting state on the physical properties of JJ. In this chapter, we also provide a modern aspects for the dependence of the supercurrent I_S on the phase difference ϕ and to discuss the forms this dependence takes in JJ of different types: superconductor-normal-superconductor (SNS), superconductor-insulator-superconductor (SIS), double barrier (SINIS), superconductor-ferromagnet-superconductor (SFS), and superconductor two-dimensional electron gas superconductor (S-2DEG-S) junctions, and superconductor-constriction-superconductor (ScS) point contacts. Unconventional symmetry in the order parameter of a high-T_c superconductor, as manifested in the CPR, will also discussed. Dynamical properties of JJ and the influence of anharmonic effects of CPR on Josephson dynamics are considered in Chap. 1. Finally, macroscopic quantum dynamics of JJ (escape rate in JJ at low temperatures, influence of multigap and d-wave symmetry of order parameter and effects of Coulomb blockade in nano-size JJ) included in this chapter.

Chapter 2 contains results of recent achievement of analog superconductivity electronics. First, we present results of experimental and theoretical study of superconducting edge bolometers. Second section includes description dynamics, time resolution and sensitivity of Josephson single junction and balanced comparators. Furthermore, we discuss single junction and double junction interfeometers and superconducting quantum interference devices (SQUIDs) on these bases. Influence of the d-wave order parameter symmetry on the flux quantization in a superconducting ring is discussed. Ultimate characteristics of alternating current (AC) and direct current (DC) SQUIDs are presented. Next section in this chapter deals with superconducting microwave devices and metrological applications. At the end of the Chap. 2, a short description of the multi-terminal devices is included.

Chapter 3 contains results of recent achievement of digital superconductivity electronics. Energy efficiency is one of the most important parameters in electronic systems, where the efficiency and performance of computer systems depends on the consumption of time and usage of resources compared to accomplished useful work. As the necessity of high performance computers, especially for research and development in various fields, continuously increases; the last two decades have shown us that straightforward use of conventional complementary metal oxide semiconductor (CMOS) logic circuits are encountering a fundamental obstacle in terms of power consumption. For the future high performance computer systems, power consumption is the main barrier in CMOS technology. At the same time, resistive single flux quantum (RSFQ) based circuits are recognized as the next generation very-large-scale integration technology by providing low energy barrier between states, which implies high operating speed with low power consumption. The main goal of the introduction part of this chapter is the description of foundations of digital superconducting electronics. Second section devoted to latching Josephson logic. Given basic gates of this logic and described main parameters. Third section deals with description of RSFQ logic circuits. RSFQ electronics is a digital logic family based on the transfer and storage of single flux quantum (SFQ). JJs act as current-controlled gates and allow the implementation of any logic function for deterministic data processing. Final section of this chapter includes recent development of RSFQ logic in past few years.

In Chap. 4, we have summarized the possible applications of JJ and interferometers in the field of quantum computation. The quantum processor then performs a quantum mechanical operation on this input state in order to derive an output which is also a quantum coherent superposition. The basic element of a quantum computer is known as a qubit, which is a linear superposition of the two quantum basis states $|0\rangle$ and $|1\rangle$. For the realization of qubit operations based on JJ and their application requires the mK temperature regions. Energy spectrum of phase and charge qubits on a JJ and superconducting single junction interferometer is analyzed using Hamilton formalism. As followed from presented discussion, anharmonic character of CPR becomes important at temperatures mK and as a result anharmonicity must be taken into account in consideration of JJ qubits. Silent qubit using anharmonic character of CPR has also been discussed. At the end of this chapter, attention is paid to the three JJ qubit systems. Last sections of this chapter describe adiabatic measurements and Rabi oscillations in qubit systems.

The last chapter of the book, Chap. 5, is devoted to chaotic phenomena in JJ systems. JJ devices could be useful for ultrahigh-speed chaotic generators for applications of code generation in spread-spectrum communications and true random number generation in secure communication and encryption. From this point, the dynamics of JJs is of great importance in contemporary superconducting electronics. The ordinary chaotic systems have one positive Lyapunov exponent and this exponent helps to mask the message. In this chapter, first we discuss physical foundation and characteristics of chaotic dynamics of single JJ with generalized CPR, including anharmonic effects. Influence of control parameters on the dynamics of fractal JJ are discussed in second section. In the next section of this chapter, chaotic dynamics of resistively coupled two JJ systems is presented. We further explore the system parameters and focus on the regions where a higher complexity is encountered. In Sect. 4 it will be shown that a simpler DC-driven JJ device can generate a wide region of hyperchaos without using an AC source compared to other studies existing in the literature. Last section related to discussion of synchronization of chaos in system of JJ with resistive coupling.

Ankara, Turkey
August 2016

Iman Askerzade
Ali Bozbey
Mehmet Cantürk

Acknowledgements

We thank Profs. F.M. Hashimzade, R.R. Guseinov, A. Gencer, E. Kurt, A. Fujimaki, M. Fardmanesh for the useful discussions, Iman Askerzade also grateful to the Abdus Salam ICTP for the hospitality during stay as an associate member. Ali Bozbey thanks to his MS students M. Eren Çelik, Murat Özer, and Yiğit Tükel for conducting the experiments and simulations results of which are shown in Chap. 3.

The circuits shown in Chap. 3 were fabricated in the clean room for analog-digital superconductivity (CRAVITY) of National Institute of Advanced Industrial Science and Technology (AIST) with the standard process 2 (STP2). The AIST-STP2 is based on the Nb circuit fabrication process developed in International Superconductivity Technology Center (ISTEC).

This work was partly supported by the TUBITAK research grant No. 111E191.

Contents

About the Authors

Iman Askerzade received the B.S. and M.S. degrees in theory of oscillation from Moscow State University, Moscow, Russia, in 1985 and the Ph.D. and Dr. Sc. degrees in condensed matter physics from the Azerbaijan National Academy of Sciences, Baku, Azerbaijan, in 1995 and 2004, respectively. He is currently a Professor with the Department of Computer Engineering, Ankara University, Ankara, Turkey, and a Principial Scientific Researcher with the Institute of Physics, Azerbaijan National Academy of Sciences. He was an Associate Member with Abdus Salam International Center for Theoretical Physics, Trieste, Italy, a Postdoctoral Researcher with Solid State Institute, Dresden, Germany (2000) and Bilkent University (2001–2002), Turkey. He is founding member of the Ankara University Superconductivity Technologies Application and Research Center (CESUR). His current research interests include computational condensed matter physics, fuzzy logic, and quantum computing.

Ali Bozbey received the B.S., M.S., and Ph.D. degrees in electrical and electronics engineering from Bilkent University, Ankara, Turkey, in 2001, 2003, and 2006, respectively. In 2002, he was a Guest Researcher with the Jülich Research Center, Jülich, Germany, and in 2007, he was a Postdoctoral Researcher with Nagoya University, Nagoya, Japan. Since 2008, he has been with the Department of Electrical and Electronics Engineering, TOBB University of Economics and Technology (TOBB ETU), Ankara, where he teaches in the areas of semiconductor and superconductor electronics. He is the head of the TOBB ETU Superconductivity Electronics Laboratory and founding member of the Ankara University Superconductivity Technologies Application and Research Center (CESUR). He acted as an executive board member at CESUR between 2012 and 2016. His research interests include design, modeling, and applications of superconductor sensors and integrated circuits.

Mehmet Cantürk received the B.S. degree in science education in physics, the M.S. degree in physics, and the Ph.D. degree in physics and the M.S. degree in information systems from Middle East Technical University, Ankara, Turkey, in 1995, 1997, and 2004, respectively. He has an experience in the curriculum development for undergraduate programs in the field of information and communication technology. He is currently a member Center of Excellence of Superconductivity Research, Ankara. His research interests include quantum computing based on superconducting qubits, embedded system development, superconducting device modeling for chaotic signal generations, and nanoscale device modeling through small-world network topology.

Chapter 1
Foundations of Josephson Junction Dynamics

Abstract The first chapter of the book is devoted to the basics of Josephson effect. Calculations of Josephson current using different approaches are presented. Results of anisotropy and multiband effects of superconducting state on Josephson current presented in second section. Experimental fabrication parameters of Josephson junctions (JJs) and study their current-phase relation (CPR) are discussed in the third section. Dynamical properties of JJ comes next. Influence of anharmonic CPR on Josephson dynamics included in fifth section. The last section of this chapter deals with macroscopic quantum effects in JJs.

1.1 Josephson Current

1.1.1 The Josephson Current Using Method of Tunnel Hamiltonian

1.1.1.1 Calculation of Current Without External Magnetic Field

In 1962 year Josephson predicted (Josephson 1962) that a supercurrent I_S could exist between two superconductors separated by a thin insulating layer (Fig. 1.1). We will now calculate the Josephson current for a tunnel junction geometry using low order perturbation theory. In fact, in superconducting state the electron operators have not been influenced by charge, which means an occupation of an effectively neutral state. Due to that a tunneling event means a real transfer of an electron, charge has to be accounted for properly. We will use electron operators for the description of the superconducting states, but charge transfer will be accounted for separately.

When an electron tunnels through the barrier, an electron and hole state is created on the opposite (left and right) side of the barrier. The corresponding Hamiltonian for this process can be written as

© Springer International Publishing AG 2017
I. Askerzade et al., *Modern Aspects of Josephson Dynamics and Superconductivity Electronics*, Mathematical Engineering, DOI 10.1007/978-3-319-48433-4_1

Fig. 1.1 Schematic structure of tunnel JJ

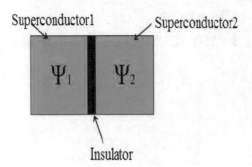

$$H = H_1 + H_2 + T, \tag{1.1}$$

$$T = \sum_{\mathbf{pq}\sigma} (T_{\mathbf{pq}} a^+_{\mathbf{p}\sigma} a_{\mathbf{q}\sigma} + T^*_{\mathbf{pq}} a^+_{\mathbf{q}\sigma} a_{\mathbf{p}\sigma}), \tag{1.2}$$

where $T_{\mathbf{pq}}$ is the tunneling matrix element, which is determine probability of tunneling. $a^+_{\mathbf{p}\sigma}$, $a_{\mathbf{p}\sigma}$ are the creation and annihilation operators of the left electrons with momentum $\mathbf{p}$ and spin σ; $a^+_{\mathbf{q}\sigma}$, $a_{\mathbf{q}\sigma}$ are the corresponding operators of the right electrons with momentum $\mathbf{q}$ and spin σ. Influence of tunneling operator T will be taken into account in framework of perturbation theory. As will be seen below, the detail behavior of the tunneling matrix element $T_{\mathbf{pq}}$ is not important. Reduced Hamiltonians of superconductors can be (for left for example) (Bogoliubov 1958; Bogolyubov 1972) written as

$$H_i = \sum_{\mathbf{p}\sigma} \frac{p^2}{2m} a^+_{\mathbf{p}\sigma} a_{\mathbf{p}\sigma} + \frac{1}{2} \sum_{\mathbf{p}\sigma} \Delta_{\mathbf{p},\sigma} (a^+_{\mathbf{p},\sigma} a^+_{-\mathbf{p},\sigma-} + c.c.) \; \forall \, i = 1, 2 \tag{1.3}$$

where $\Delta_{\mathbf{p},\sigma}$ denotes that superconducting order parameter in electrode. The total current from left electrode to right electrode is given the transition rate

$$I = -e \left\langle \frac{dN_1}{dt} \right\rangle = -e \sum_{\mathbf{p},\sigma} \left\langle \frac{dn_{\mathbf{p},\sigma}}{dt} \right\rangle, \tag{1.4}$$

where N_1 is number of left electrons, $\langle .... \rangle$ means the expectation value in equilibrium state. Heisenberg equation for left electrons has a form

$$i \frac{da_{\mathbf{p}\sigma}}{dt} = \frac{p^2}{2m} a_{\mathbf{p}\sigma} + \Delta_{\mathbf{p},\sigma} a^+_{-\mathbf{p},\sigma-} + \sum_{\mathbf{q}} T_{\mathbf{pq}} a_{\mathbf{q}\sigma}. \tag{1.5}$$

Using last and conjugated equation, we can get

$$\left\langle \frac{dn_{\mathbf{p}\sigma}}{dt} \right\rangle = 2 \, \mathrm{Im} \left\{ \sum_{\mathbf{q}} T_{\mathbf{pq}} \langle a^+_{\mathbf{p}\sigma}(t) a_{\mathbf{q}\sigma}(t) \rangle - \Delta^*_{\mathbf{p},\sigma} \langle a_{-\mathbf{p},-\sigma}(t) a_{\mathbf{p},\sigma}(t) \rangle \right\}. \tag{1.6}$$

Using phase presentation of operators of left and right superconductors

$$a_{\mathbf{p}\sigma}(t) = \tilde{a}_{\mathbf{p}\sigma}(t)\exp(-i\phi_1(t)/2); \quad a_{\mathbf{q}\sigma}(t) = \tilde{a}_{\mathbf{q}\sigma}(t)\exp(-i\phi_2(t)/2) \qquad (1.7)$$

and fact that imaginary part of second term in Eq. (1.6) is zero, we can get

$$\left\langle \frac{dn_{\mathbf{p}\sigma}}{dt} \right\rangle = 2\,\mathrm{Im}\left\{ \sum_{\mathbf{q}} T_{\mathbf{pq}} \langle \tilde{a}_{\mathbf{p}\sigma}^+(t)\tilde{a}_{\mathbf{q}\sigma}(t) \rangle \exp(-i(\phi_1(t) - \phi_2(t))/2) \right\}. \qquad (1.8)$$

Final expression for supercurrent has a form

$$I = I_c \sin\phi; \quad \phi = \phi_1 - \phi_2, \qquad (1.9)$$

where critical current I_c is determined as

$$I_c = 4eN_1(0)N_2(0)\langle T^2\rangle \int_{-\infty}^{\infty}\int_{-\infty}^{\infty} T\sum_{\omega} \frac{\Delta_1\Delta_2 d\xi_p d\xi_q}{(\omega^2 + \xi_p^2)(\omega^2 + \xi_q^2)}. \qquad (1.10)$$

Average value of tunneling matrix element in Eq. (1.10) over Fermi surfaces of both superconductors can be written as

$$\langle T^2\rangle = V_1 V_2 \int \frac{d\Omega_1}{4\pi} \int \frac{d\Omega_2}{4\pi} |T_{\mathbf{pq}}|^2, \qquad (1.11)$$

where V_1 and V_2 is are the volumes of Fermi sphere of corresponding supercon-ductors. It is convenient to express the tunneling matrix element in terms of the normal-state resistance of the junction, obtained from relation

$$eR_N = \frac{1}{eN_1(0)N_2(0)\langle T^2\rangle} \qquad (1.12)$$

where $N_i(0)$ denotes density of state at the Fermi level of superconducting compounds ($i = 1, 2$). Critical current in the case of the same superconducting order parameters in left and right side ($\Delta_1 = \Delta_2 = \Delta$) has a form (Fig. 1.1),

$$I_c = \frac{\pi\Delta}{2eR_N}\tanh\frac{\Delta}{2T}, \qquad (1.13)$$

while for different superconductors at $T = 0$ ($\Delta_1 \neq \Delta_2$) is true the relation

$$I_c = \frac{2\Delta_1\Delta_2}{eR_N(\Delta_1 + \Delta_2)} K\left(\frac{|\Delta_1 - \Delta_2|}{(\Delta_1 + \Delta_2)}\right), \qquad (1.14)$$

where $K\left(\frac{|\Delta_1-\Delta_2|}{(\Delta_1+\Delta_2)}\right)$ elliptical integral of I type. This equation is well known as Ambegaokar–Baratoff result for critical current in tunnel JJ (Ambegaokar and Baratoff 1963).

It is useful to note that a dissipative mechanism quasiparticle tunneling produces decoherence. Although it is shown to be absent for subgap voltage measurements of real Josephson junctions reveal a small subgap current. This current originates from multiple Andreev reflections (Andreev 1964), can be considered as as higher-order tunneling processes. Hence a description of the tunnel junction that easily predicts the processes for arbitrary tunneling matrix elements is needed. This is especially required since real tunnel junctions do not have constant matrix elements.

1.1.1.2   Magnetic Field Dependence of Critical Current

The first characteristic of JJ is product of $I_c R_N$ which should be of order $2\Delta/e$ independently device geometry. The second important property of JJ is the oscillatory characteristic of maximal critical current versus applied magnetic field perpendicular to the flow of the supercurrent. In the presence of external magnetic field in junction plane (Fig. 1.1) the Josephson phase ϕ becomes function of coordinates. If cross section of the junction has rectangular form with area S and penetration depth of magnetic field is δ, then additional phase jump due to magnetic field has a form

$$\frac{\partial\phi}{\partial x} = 4e\delta H_y. \tag{1.15}$$

Corresponding density of Josephson current in presence external magnetic field $\mathbf{H} = (0, H, 0)$ calculated as

$$j(x) = j_{max} \sin(\phi_0 + 4e\delta H_y x). \tag{1.16}$$

Total current throught JJ can written as result of averaging procedure

$$I = \frac{1}{L}\int_0^L j(x)dx = j_{max}\frac{\cos\phi_0 - \cos(\phi_0 + 4e\delta H_y L)}{4e\delta H_y L}. \tag{1.17}$$

Maximal current correspond to maximum of the last expression and has a form (Rowell 1963)

$$I_c(H) = I_c(0)\left|\frac{\sin(\frac{\pi\Phi}{\Phi_0})}{\frac{\pi\Phi}{\Phi_0}}\right|. \tag{1.18}$$

As followed from last equation, critical current of JJ in external magnetic fields is equal to zero at $\Phi = n\Phi_0$, where $\Phi_0 = \frac{h}{2e}$ is the flux quantum (Fig. 1.2). In other words oscillatory response of the supercurrent to the magnetic field B reveal Fraunhofer-like diffraction pattern. This property of the JJ make this devices high

Fig. 1.2 Magnetic field dependence of critical current, I_c

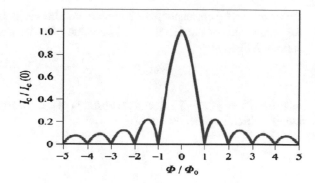

sensitive switching elements in superconductivity electronics at the level of Φ_0 (Likharev 1986; Barone and Paterno 1982).

1.1.2 Josephson Current Using GL Theory

By further studies have shown (see for example Likharev 1979) that, the effect extends beyond Josephson's predictions and can exist if superconductors are connected by a "weak link" of any physical nature (normal metal, semiconductor, superconductor with smaller critical temperature, geometrical constriction, etc. Fig. 1.3). For the calculation of supercurrent in structures type ScS ($L \ll \xi$) Aslamazov and Larkin (1969) used Ginzburg–Landau (GL) equation. At temperatures near the critical temperature T_c, a superconductor is described by the GL equation (Ginzburg and Landau 1950; Tinkham 1996)

$$-\xi^2 \nabla^2 \psi - \psi + \psi^3 = 0, \qquad (1.19)$$

where $\psi = \frac{\Psi}{\Psi_0}$ normalized superconducting order parameter. Equation (1.19) can be derived directly from the Bardeen–Cooper–Schrieffer (BCS) equation using expansion over small gradients and substituting it into the self-consistency equation (Abrikosov 1988; Abrikosov et al. 1975). The GL equation describe the spatial evolution of ψ at a length scale of the order of $\xi^2(T)$. They should be completed at

Fig. 1.3 Schematic structure of JJ ScS type

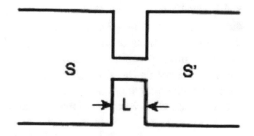

the interfaces by corresponding boundary conditions, knowledge of which is crucial
for application of the GL theory. The value of order parameter in the equilibrium
state is defined as

$$|\Psi_0|^2 = -\frac{2\alpha(T)}{\beta} = \frac{2\gamma(T_c - T)}{\beta}, \quad (1.20)$$

where $\alpha(T) = \gamma(T_c - T)$, γ and β constants of GL theory. $\xi(T)$ is the GL coherence
length, which is given by

$$\xi^2(T) = \frac{\hbar^2}{4m\gamma(T_c - T)}. \quad (1.21)$$

The supercurrent in the GL theory is given by the expression (Aslamazov and Larkin
1969)

$$I = \frac{\alpha\hbar e}{\beta m} \text{Im}(\psi^*\psi) = I_c \sin\phi; \; I_c = \frac{\alpha\hbar e}{\beta m}. \quad (1.22)$$

Presented theory is applicable to temperatures close to T_c, and it has been commonly used to describe current-carrying states in various types of JJs because of its
computational simplicity compared to the quasiclassical methods. Additionally, the
GL theory is physically more clear, since superconducting state is described by a
single function ψ, which corresponds to the superconducting order parameter. It is
also commonly referred to as a condensate wave function.

This book deals with the recent developments of Josephson dynamics and superconductivity electronics. Discoveries of new classes of superconductors and their
experimental investigations revealed a great variety its physical properties.

Equation (1.22) is very general: it does not depend on the electronic mean free
path in the weak link and is applicable to all types of weak links near T_c. According
to Eq. (1.22), near T_c the CPR of the junction is always sinusoidal, independent of
the material of the weak link, which is characterized only by its resistance R_N.

1.1.3 Josephson Current Using Bogoliubov-de Gennes Equations

Kulik and Omelyanchuk (1975) considered a ScS JJs as a diffusive quasi-one-
dimensional wire (Fig. 1.3), connecting two superconductors and used Usadel equations for calculations of Josephson current (Usadel 1970; Golubov et al. 2004). In
the dirty limit superconductor is described by the Usadel equations for normal and
anomalous Green's functions (Usadel 1970). For a symmetric junction formed by
identical superconductors ($\Delta_1 = \Delta_2 = \Delta$) result of calculation of Josephson current
can be presented as Kulik and Omelyanchuk (1975)

$$I(\phi) = \frac{4\pi T}{eR_N} \sum_{\omega>0} \frac{\Delta\cos(\phi/2)}{\delta} \arctan\left(\frac{\Delta\sin(\phi/2)}{\delta}\right), \quad (1.23)$$

where $\delta = \sqrt{\omega_n^2 + \Delta^2 \cos^2(\phi/2)}$ and $\omega_n = \pi kT(2n+1)$ are the Matsubara frequency (Abrikosov et al. 1975). The curves $I(\phi)$ are nonsinusoidal at low temperatures and reduce to the Aslamazov–Larkin (1969) result for $T \to T_c$.

Further, Kulik and Omelyanchuk (1978) advanced a theory, to describe the Josephson effect in the fully ballistic case. The supercurrent was calculated in the framework of the Eilenberger equations (Eilenberger 1968). The junction may contain a single scattering center, e.g., a tunnel barrier of finite transparency, characterized by the transmission coefficient D_0, where $0 < D_0 < 1$. As mentioned above the fully ballistic case means $D_0 = 1$. Corresponding result for supercurrent is (Kulik and Omelyanchuk 1978)

$$I(\phi) = \frac{\pi \Delta}{eR_N} \sin(\phi/2) \tanh\left(\frac{\Delta \cos(\phi/2)}{2T}\right). \tag{1.24}$$

Central moment to understanding the Josephson effect in such Josephson structures will be the quasiparticle bound state (Bratus et al. 1997; Kulik and Yanson 1972). To understand how to calculate a bound state, we will first consider a normal-metal tunnel junction and with a δ–function barrier $V\delta(x)$, as illustrated in Fig. 1.4. A superconducting junction SNS will also have bound states of quasiparticles excitations. These bound states describe the Josephson effect since virtual quasiparticle tunneling was necessary for the perturbation calculation. The Bogoliubov-de Gennes equations describe the spatial two-component wave-functions $\{u(\mathbf{r}), v(\mathbf{r})\}$, whose eigenstates are given by the solution of the system equations (Svidzinski 1982)

$$\left\{\frac{\hat{p}^2}{2m} - \mu\right\} u(\mathbf{r}) - \Delta(\mathbf{r})v(\mathbf{r}) = \varepsilon u(\mathbf{r}), \tag{1.25}$$

$$\left\{\frac{\hat{p}^2}{2m} - \mu\right\} v(\mathbf{r}) + \Delta^*(\mathbf{r})u(\mathbf{r}) = -\varepsilon v(\mathbf{r}), \tag{1.26}$$

where $\hat{p}$ is momentum operator and μ is chemical potential.

We can solve for the quasiparticle bound states using Eqs. (1.25) and (1.26). An incoming quasiparticle state, point A in Fig. 1.5, is reflected of the tunnel barrier to states B and C and is transmitted to states D and E. The energies of the quasiparticle bound states are

$$E_J = \Delta\sqrt{1 - D_0 \sin^2(\phi/2)}. \tag{1.27}$$

Fig. 1.4 Model of JJ with delta potential

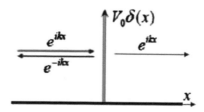

Fig. 1.5 Bound states of
quasiparticles in JJ

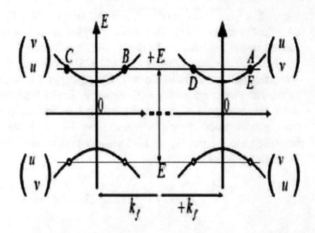

Because these two states have energies less than the gap energy E_J, they are energet-
ically bound to the junction and thus have wavefunctions that are localized around
the junction. The dependence of the quasiparticle bound-state energies on junction
phase is plotted in Fig. 1.6 for several values of transmission coefficient D_0. The
ground state is normally filled, similar to the filling of quasiparticle states of negative
energy. The current of the each bound state is given by the derivation of its energy

$$I(\phi) = \frac{2\pi}{\Phi_0} \frac{\partial E_J}{\partial \phi}. \tag{1.28}$$

A multichannel generalization of last equations was performed in Beenakker
(1991)

$$I(\phi) = \frac{e\Delta^2}{2\hbar} \sin\phi \sum_{n=1}^{N} \frac{D_n}{E_n} \tanh \frac{E_n}{2T}, \tag{1.29}$$

Fig. 1.6 Bound state energy
as function of phase for
different transmission
coefficient D_0

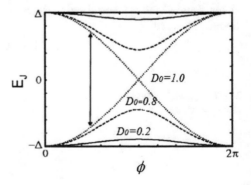

where D_n are the eigenvalues of an arbitrary transmission matrix that describes a disordered junction, n is the number of channels, and E_n is the energy of the Andreev bound state (see Eq. 1.27). Detail analysis of Josephson current in dirty diffusive junctions presented in review (Golubov et al. 2004).

1.2 Influence of Anisotropy and Multiband Effects of Superconducting State on Josephson Current

1.2.1 JJs on D-Wave Superconductors

For the calculation of Josephson current in d-wave structures it is easy to use a GL theory of d-wave superconducting state based on its symmetry properties. It is well known that d-wave order parameter symmetry was observed in cuprate high -T_c superconductors (HTS) (Kirtley 2010; Sigrist and Rice 1992; Tsuei and Kirtley 2000; Askerzade 2012). For this symmetry we use a complex order parameter, which behaves the same way as the pair wave function in momentum space

$$\psi(\mathbf{k}) = \cos k_x - \cos k_y. \tag{1.30}$$

The GL free-energy functional F for d-wave superconductors has a form

$$F = \int dV \left\{ A(T)\psi^2 + \beta\psi^4 + K_1(|D_x\psi|^2 + |D_y\psi|^2) + K_2(|D_z\psi|^2) \right\} + \frac{1}{8\pi}\mathbf{B}(\mathbf{B} - 2\mathbf{H}). \tag{1.31}$$

In Eq. (1.31) the real coefficients β and K_i are phenomenological parameters, and $A(T) = \gamma(T - T_c)$ changes sign at the superconducting transition temperature T_c. The symbols D denote the components of the gauge-invariant gradient $D = (\nabla - \frac{2\pi i \mathbf{A}}{\Phi_0})$, where $\mathbf{A}$ is the vector potential ($\mathbf{B} = curl\mathbf{A}$). For the calculation of the Josephson current, it is useful to introduce the coupling between the order parameters of two linked superconductors (Fig. 1.7). Coupling between superconductors can be expressed by

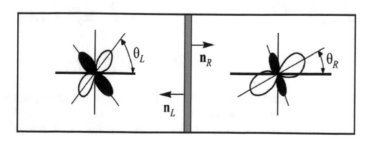

Fig. 1.7 Schematic diagram of JJ based on d-wave superconductors

adding term

$$F = t \int dS \chi_1(\mathbf{n}_1) \chi_2(\mathbf{n}_2) \left\{ \psi_1^* \psi_2 + \psi_2^* \psi_1 \right\}, \tag{1.32}$$

where t is a real parameter denoting the coupling strength. The functions $\chi_j(\mathbf{n}_j)$ are symmetry functions of the interface normal vector $\mathbf{n}_j$ in the crystal basis of the side (j). For the current density perpendicular to the interface we can get

$$I(\phi) = \frac{4\pi c t}{\Phi_0} \chi_1(\mathbf{n}_1) \chi_2(\mathbf{n}_2) \, |\psi_1| \, |\psi_2| \sin \phi. \tag{1.33}$$

For the d-wave superconductors using symmetry function as $\chi(\mathbf{n}) = n_x^2 - n_y^2$ leads to final Sigrist–Rice result for clean JJ (Sigrist and Rice 1992)

$$I(\phi) = A_s \cos(2\theta_L) \cos(2\theta_R) \sin \phi, \tag{1.34}$$

where θ_L, θ_R the angles of the crystallographic axes with respect to the interface, A_s is a constant characteristic of the junction (Fig. 1.7) (Geshkenbein et al. 1987). For the dirty limit of JJ

$$I(\phi) = A_s \cos 2(\theta_L + \theta_R) \sin \phi. \tag{1.35}$$

Based on a Green's function method, the Josephson current in a d-wave/ insulator/ d-wave superconductor $(d/I/d)$ junction is calculated by taking into account the anisotropy of the pair potentials explicitly (Tanaka 1994; Tanaka and Kashiwaya 1997)

$$R_N I(\phi) = \frac{\pi \bar{R}_N kT}{e} \left\{ \sum_{\omega_n} \int_{-\pi/2}^{\pi/2} \left[\frac{a_1(\theta, i\omega_n, \phi)}{\Omega_{L,+}} |\Delta_L(\theta_+)| - \frac{\tilde{a}_1(\theta, i\omega_n, \phi)}{\Omega_{L,-}} |\Delta_L(\theta_-)| \right] \cos \theta d\theta \right\}, \tag{1.36}$$

where $\Omega_{n,L,\pm} = \sqrt{\Delta_L^2(\theta_\pm) + \omega_n^2}$. The quantity R_N denotes the normal resistance and $\bar{R}_N$ is expressed as

$$\bar{R}_N^{-1} = \int_{-\pi/2}^{\pi/2} \sigma_N \cos \theta d\theta; \quad \sigma_N = \frac{4Z_0^2}{(1 - Z_0^2) \sinh^2(\lambda d_i) + 4Z_0^2 \cosh^2(\lambda d_i)}, \tag{1.37}$$

$$\lambda = \sqrt{1 - \kappa^2 \cos^2 \theta} \lambda_0, \quad Z_0 = \frac{\kappa \cos \theta}{\sqrt{1 - \kappa^2 \cos^2 \theta}}. \tag{1.38}$$

Here, σ_N denotes the tunneling conductance for the injected quasiparticle when the junction is in the normal state. The Andreev reflection coefficient $a_1(\theta, i\omega_n, \phi)$ is obtained by solving the Bogoliubov-de Gennes equation, and $\tilde{a}_1(\theta, i\omega_n, \phi)$ is obtained by substituting $\pi - \theta$, $-\phi_L$, and $-\phi_R$ for θ, ϕ_L, and ϕ_R into $\tilde{a}_1^\sim(\theta, i\omega_n, \phi)$

respectively. If we take only the $\theta = 0$ component, the magnitude of the Josephson current is proportional to $\cos(2\theta_L)\cos(2\theta_R)$, and the phenomenological theory by Sigrist and Rice (1992) is reproduced.

In general, supercurrent in JJ with d-wave superconductors $I(\phi)$ can be decomposed into the series of $sin(n\phi)$ and $cos(n\phi)$

$$I(\phi) = \sum_{n>1}(I_n \sin n\phi + J_n \cos n\phi). \tag{1.39}$$

The last expression contains the component of the Josephson current carried by the multiple reflection process at the interface of junction. In Eq. (1.39), the n th current components correspond to the amplitudes of the n th reflection processes of quasiparticles. For $\sigma_N \sim 0$, Josephson current, $I(\phi)$ is sinusoidal and the results of Ambegaokar-Baratoff theory (Ambegaokar and Baratoff 1963) is reproduced. When $\sigma_N = 1$, Eq. (1.39) reproduces the classical results by Kulik and Omelyanchuk (Kulik and Omelyanchuk 1975, 1978). On the other hand as followed from Tanaka analysis (Tanaka 1994; Tanaka and Kashiwaya 1997), for a fixed Josephson phase difference, depending on the injection angle of the quasiparticle, the component of the supercurrent becomes either positive or negative. In some cases, the phase difference ϕ_0, which corresponds to the free energy minima, is located at neither zero nor π. When the crystal axis in superconducting electrodes is tilted from the interface normal of Josephson Junction, mid-gap states are formed near the interface, which depends on the angle of the crystal axis and the injection angle of the quasiparticle. This effects leads to the enhancement of the Josephson current at low temperatures.

1.2.2 JJ Between Two-Band S-Wave Superconductors

After the discovery of superconductivity in MgB_2 with $T_c = 39$ K (Nagamatsu et al. 2001), multiband superconductivity became actual topic in solid state physics. It's well known that pairing mechanism in MgB_2 has electron-phonon origin and order parameters have s-wave symmetry. Nonmagnetic borocarbides and Fe-based superconductors, which have discovered recently (Askerzade 2012), Izyumov and Kurmaev (2010) can be adding on to multiband compounds. In this section we present results the stationary Josephson effect in ScS junction based on two-band superconductors. The presented microscopic theory of the «dirty» S-c-S junction for two-band superconductors, generalize the Kulik–Omelyanchouk theory for this case (Yerin and Omelyanchouk 2010). In this section the case of dirty two-band superconductor with strong impurity intraband scattering rates and weak interband scattering is presented (Fig. 1.8).

As mentioned in Yerin and Omelyanchouk (2010), in the dirty limit superconductor is described by the Usadel equations for normal and anomalous Green's functions.

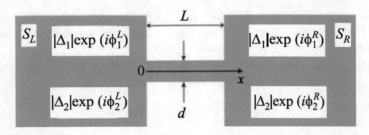

Fig. 1.8 Schematic diagram of JJ based on two-band superconductors

Calculation Josephson current without interband scattering between two-band super-conductors leads to Yerin and Omelyanchouk (2010),

$$I(\phi) = \frac{4\pi T}{e} \sum_{i=1}^{2} \frac{1}{eR_{N_i}} \sum_{\omega>0} \frac{\Delta_i \cos(\phi/2)}{\sqrt{\omega^2 + \Delta_i^2 \cos^2(\phi/2)}} \arctan \frac{\Delta_i \sin(\phi/2)}{\sqrt{\omega^2 + \Delta_i^2 \cos^2(\phi/2)}}.$$

(1.40)

As you can see from Eq. (1.40), current flows independently from the first (second) band to the first (second) one. This equation is straightforward generalization of Ambreokar–Baratof results to the case of two-band superconductor (Ambegaokar and Baratoff 1963). Introducing the total resistance as $R_N = R_{N1}R_{N2}/(R_{N1} + R_{N2})$ and normalizing the current on the value $I_0 = (2\pi/eR_N)T_c$ the CPR $I(\phi)$ Eq. (1.40) for different values of $r = R_{N1}/R_{N2}$ and temperature T plotted in Fig. 1.9. Using perturbations theory in the first approximation for Green's functions in each band

Fig. 1.9 CPR of JJ on two-band superconductor (MgB_2/MgB_2) for different ratios of resistances $r = 0.1; 10$ af different temperatures $T = 0(1); 0.5T_c(2); 0.9T_c(3)$

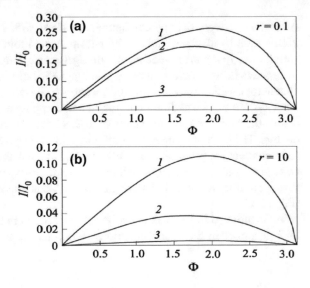

for the case of nonzero interband scattering the corrections to the current (1.40) $\delta I = \delta I_1 + \delta I_2$ were obtained as

$$\delta I_1 = \frac{2\pi T \gamma_{12}}{eR_{N1}} \sum_{\omega} \left\{ \frac{\omega^2(\Delta_2 e^{i\delta} - \Delta_1)\cos(\phi/2)}{\sqrt{(\omega^2 + \Delta_1^2 \cos^2(\phi/2))^3}\sqrt{(\omega^2 + \Delta_2^2)}} \arctan \frac{\Delta_1 \sin(\phi/2)}{\sqrt{\omega^2 + \Delta_1^2 \cos^2(\phi/2)}} + \frac{1}{2} \frac{\omega^2(\Delta_2 e^{i\delta} - \Delta_1)\sin(\phi)}{(\omega^2 + \Delta_1^2)\sqrt{(\omega^2 + \Delta_2^2)(\omega^2 + \Delta_1^2 \cos^2(\phi/2))}} \right\},$$

(1.41)

$$\delta I_2 = \frac{2\pi T \gamma_{21}}{eR_{N2}} \sum_{\omega} \left\{ \frac{\omega^2(\Delta_1 e^{i\delta} - \Delta_2)\cos(\phi/2)}{\sqrt{(\omega^2 + \Delta_2^2 \cos^2(\phi/2))^3}\sqrt{(\omega^2 + \Delta_1^2)}} \arctan \frac{\Delta_2 \sin(\phi/2)}{\sqrt{\omega^2 + \Delta_2^2 \cos^2(\phi/2)}} + \frac{1}{2} \frac{\Delta\omega^2(\Delta_2 e^{i\delta} - \Delta_1)\sin(\phi)}{(\omega^2 + \Delta_2^2)\sqrt{(\omega^2 + \Delta_1^2)(\omega^2 + \Delta_2^2 \cos^2(\phi/2))}} \right\},$$

(1.42)

where γ_{12} parameter of interband scattering, δ-shift of phase. When the interband scattering is taken into account and if phase shift $\delta \neq 0$, the phases of Green functions ϕ_i not coincide with phases of order parameters Δ_i. Temperature dependences of critical current $I_c(T)$ for different values of r presented in Fig. 1.10.

It is useful to note that the extended s-wave or $s\pm$ pairing symmetry with a π phase shift between the hole and electron Fermi surface has attracted considerable attention and is favored by many experiments and is considered as a most probable candidate for Fe based superconductors (Askerzade 2012; Mazin et al. 2008). Generalization of of above presented theory to the case $s\pm/I/s$ junctions presented in study Lin (2012). For this junctions there are exists frustrated interaction among different superconducting order parameters, and under appropriate conditions, time-reversal symmetry is broken. Depending on the competition of interelectrode Josephson couplings and interband Josephson coupling, the interference between the two tunneling channels can change continuously from adding constructively, where they have the same phase to canceling destructively where they have a π phase shift.

Fig. 1.10 Temperature dependence of critical current of TB JJ for different ratios of resistances $r = 0.1(1); 1(2); 10(3)$

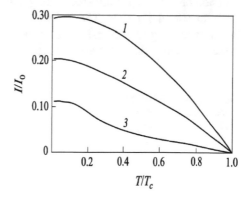

1.2.3 CPR of S/F/S and S/AF/S Structures

In this section, we concentrate our attention on the approach based on the Usadel equation and consider the $S/F/S$ junction with a ferromagnetic (F) layer of thickness $2d_r$ (Fig. 1.11).

In calculation of supercurrent, we use the following equation

$$I(\phi) = ieN(0)D_f \pi TS \sum_{-\infty}^{\infty} \left(F\frac{d\tilde{F}}{dx} - \tilde{F}\frac{dF}{dx} \right), \tag{1.43}$$

where Green functions are satisfied the condition $\tilde{F}(x, h) = F^*(x, -h)$ and determined by the expressions: $F_s(-d_r) = \frac{\Delta e^{-i\phi/2}}{\sqrt{\Delta^2 + \omega^2}}$, $F_s(d_r) = \frac{\Delta e^{\phi/2}}{\sqrt{\Delta^2 + \omega^2}}$, S is the area of the cross section of the junction, D_f is the diffusion coefficient in F layer, and $N(0)$ is the electron density of state for a one-spin projection. Calculations using Eq. (1.43) gives the usual sinusoidal CPR with the critical current (Buzdin 2003)

$$I_c = eN(0)D_f \pi TS \sum_{-\infty}^{\infty} \left(\frac{\Delta^2}{\omega^2} \frac{2k/\cosh(2kd_r)}{\tanh(2kd_r)(1 + \Gamma_\omega^2 k^2) + 2k\Gamma_\omega} \right), \tag{1.44}$$

where $\Gamma_\omega = \frac{\gamma_B \xi_N}{G_s}$. Generalization of preceding theory to the case of the different interface transparencies presented in article (Buzdin and Kupriyanov 1991)

$$I_c = \frac{V_0}{R_N} 4y \frac{\cos 2y \sinh(2y) + \sin 2y \cosh(2y)}{\cosh(4y) - \cos 4y}), \tag{1.45}$$

where introduced a new notation $y = \frac{d_F}{\xi_F}\sqrt{\frac{H_0}{2\pi T_c}}$, H_0 is the exchange energy in F layer. The S/F/S junctions reveal the nonmonotonic behavior of the critical current as a function of the F layer thickness. The point at which $I_c = 0$ signals the transition from the 0 to the π state. The first vanishing critical current occurs at $2y_c = 2.36$ which is exactly the critical value of the F layer thickness in the S/F multilayer system corresponding to the 0-π -state transition (Fig. 1.12) (Golubov et al. 2004).

Fig. 1.11 Structure of SFS junction (γ_B is transparency coefficient of boundary)

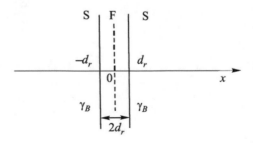

Fig. 1.12 Critical current of
I_c of SFS junction as
function $y = \frac{d_F}{\xi_F}\sqrt{\frac{H}{2\pi T_c}}$,
where H is the exchange
energy in F layer

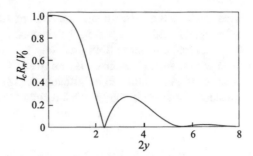

CPR in different structures as SFcFS and double-barrier SIFIS presented in
Golubov et al. (2002). The critical current of SIFIS junctions reveals nonmonotonic
temperature dependency. Deformation of CPR of double-barrier SIFIS junctions for
the different exchange integral in F layer presented in Fig. 1.13. A similar change in
CPR was experimentally observed in study (Demler et al. 1998). One of the interesting properties of SFS systems is the rotation of the magnetization vector of F
layers under an external magnetic field (Bergeret et al. 2005; Garifullin et al. 2002;
Komissinskiy et al. 2007). Details of CPR of different Josephson structures with F
layer were presented in reviews (Golubov et al. 2004; Buzdin 2005). It is well known
that the study of the CPR is also important for understanding the fundamental properties of superconducting materials, such as the symmetry of the superconducting order
parameter and peculiarities of spin transport in systems based on superconducting
and ferromagnetic materials.

Despite wide discussion of JJ with ferromagnetic F layers, there are few papers
related with investigations of structures with anti-ferromagnetic (AF) layers. Firstly
such structure was studied by Gorkov and Kresin (2002). They found that the critical
current strongly depends external magnetic field. An analytical expression can be
written as

$$I_c(M_s) = I_{co}\sqrt{\frac{2}{\pi \gamma M_s}}\cos\left(\beta M_s - \frac{\pi}{4}\right), \qquad (1.46)$$

Fig. 1.13 CPR of SIFIS for
different exchange integral
of F layer

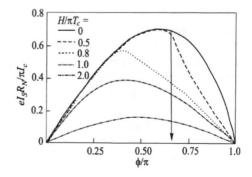

where $\gamma \gg 1$ related with characteristic of AF layer, $0 < M_s < 1$ parameter of AF ordering, I_{co} critical current in the absence of external magnetic field, which equal to the critical current in SNS junctions. It is useful to note the importance of study of JJ with magnetic ordering layers from the point application in superconducting electronics. As mentioned in Likharev (2012), such Josephson structures may allow substantial savings in the Josephson circuit area.

1.3   Fabrication Parameters of JJs

1.3.1   Important Parameters of JJs

As mentioned above, the main important parameter of JJs is the $I_c R_N$ product. It is well known that the $I_c R_N$ product determines the high frequency limit/speed and the amplitude of the output signal for the JJs. Dependency of $I_c R_N$ on the critical current density J_c is altered, it may be proportional to J_c^n, where n varies between 0.4 and 0.7 (Brake et al. 2006). An obvious way to increase the frequency limit is to increase J_c. High $I_c R_N$ can be reached in principle using HTS material. HTS JJs promise about 10 times a higher circuit speed than low temperature superconductors (LTS) (due to the 10 times larger $I_c R_N$ product of HTS) and have the additional advantage of being of the "self-shunted" type, eliminating the need for chip area-wasting external shunts. Values of 8 and 10 mV at 4 K were reached, respectively, using boundary bicrystal junctions or ramp type JJs. These results indicate that circuit operation up to 5 THz can be achieved. The investigated step-edge and bicrystal junctions routinely made by many groups had a width of weak link ranging from 1 to 50 μm and with a thickness of 0.1–0.3 μm. The scaling relation between $I_c R_N$ and J_c for both step edge and bicrystal junctions fabricated from YBCO and GdBaCuO and tested at 77 K is shown in Fig. 1.14 (Haupt et al. 1997).

1.3.1.1   Step-edge Junctions

Step-edge junctions are formed by depositing a thin HTS epitaxial film on substrate that has a step etched into the surface. The weak link is then formed as the single-layer film bridges the two levels. The steps usually are formed by patterning and ion milling the $SrTiO_3$ substrate. There are significant advantages to step-edge technology. A variety of large-area substrates can be employed in the step-edge process (bicrystal technology is limited to a 10-mm $SrTiO_3$ substrate). The step-edge height is highly reproducible. However, the junction parameters appear to depend greatly on film thickness, and variations observed for step-edge technology are probably a result of the film thickness and superconducting parameters variations, on the "bottom" and on the "top" of the step.

Fig. 1.14 Relation between
$I_c R_n$ and J_c for YBCO and
GdBCO based JJ (Haupt
et al. 1997)

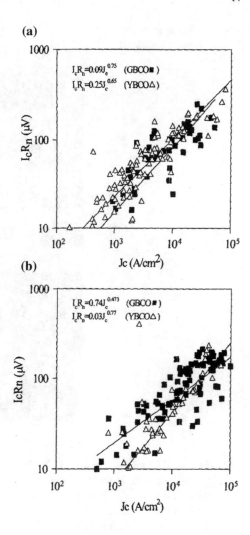

(a)

$I_c R_n = 0.09 J_c^{0.75}$ (GBCO ■)
$I_c R_n = 0.25 J_c^{0.65}$ (YBCO △)

(b)

$I_c R_n = 0.74 J_c^{0.473}$ (GBCO ■)
$I_c R_n = 0.03 J_c^{0.77}$ (YBCO △)

1.3.1.2 Bicrystal Junctions

Bicrystal junctions are fabricated on the substrate that has a twin boundary formed by two single-crystal domains with different crystal orientations. On SrTiO$_3$ bicrystal substrates, a misorientation angle usually is an order 25–37°. They are made by fusing two separate single crystals and then dicing substrates from the single piece. This results in a twin boundary down the center of the substrate. When a superconductor film is epitaxially grown, a twin boundary forms in the superconductor film at this interface.

In traditional Nb-technology complete wafer trilayer processes on the basis of SIS, SINIS and SNS sandwiches. S = Nb, I = Al-oxide, N = Al, Pd, Au, Cu, Ti, HfTi, or multilayer structures, e.g., HfTi/Nb/HfTi was used (Brake et al. 2006). For

Fig. 1.15 Cross section of
SINIS junctions (Buchholz
et al. 2001)

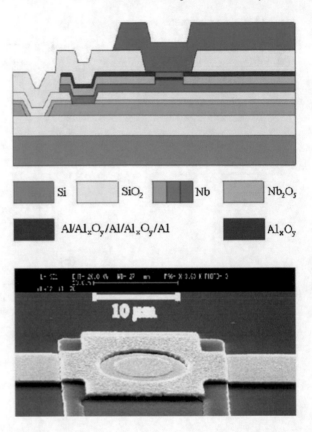

low complexity circuits (20 junctions) a minimum junction size of $0.3 \times 0.3\,\mu\text{m}^2$
has been reached by e-beam lithography. To utilize the advantageous properties of
internally shunted JJs implemented as the active circuit elements into superconductive circuit architectures, circuit fabrication last years is also focused on SINIS and
SNS technologies (Buchholz et al. 2001). The process was improved to raise the
characteristic voltage of SINIS two-tunnel JJs up to $V_c = 245\,\mu\text{V}$ (corresponding
critical current density: $J_c = 2.2\,\text{kA/cm}^2$). The area of the smallest junction is $A = 12\,\mu\text{m}^2$ (Buchholz et al. 2001). Figure 1.15 shows a schematic of the cross section
of a SINIS junction and a top view microphotograph. IV curve of such junctions is
nearly hysteresis-free.

It is useful to note that, there is noble technological process for fabricating HTS
JJs with reproducible characteristics. Very recent review of physical properties HTS
based JJ presented in Ovsyannikov and Konstantinyan (2012). At the same time,
high-quality tunnel junctions can be routinely fabricated for LTS.

1.3.2   Experimental Results of Study of CPR in Different JJs

As followed from above presented theoretical review (Sects. 1.1 and 1.2), in general case CPR in Josephson structures is determined by the types of the junctions. At high temperature $T_c - T \ll T_c$ deviation of CPR $I(\phi)$ from $\sin\phi$ is negligible for any type of JJ. At low temperatures $T \ll T_c$ relation $I(\phi) = I_c \sin\phi$ takes place for SIS LTS junctions (Dahm et al. 1968). In study Dahm et al. (1968) it was shown high accuracy of experimental verification of sinusoidal character of CPR using plasma resonance technique many years ago. Recently, Gronbech–Jensen et al. (2004) studied the dynamics of the tunnel JJs, simultaneously carrying DC and AC currents. This study measured the statistics of switching of LTS tunnel junctions of the Nb–NbAlOx–Nb type to the resistive state. The critical current statistics in this junction, which is controlled by thermal fluctuations at the bottom of the potential well $U(\phi) = E_J(-i\phi + 1 - \cos\phi)$, was determined for 10000 events. By changing the AC current amplitude, it was achieved the DC current corresponding to a peak in the distribution of switching events.

New method for CPR measurement and some of its practical applications presented in Ilichev et al. (2001a, b). Most commonly for the experimental investigation of the CPR, JJ of interest is incorporated in a superconducting ring with a sufficiently small inductance L. This circuit is usually called a single junction interferometer (Likharev 1986) in contrast to terminology in other books. Under conditions $\frac{\omega L}{R_N} \ll 1$, $\omega^2 LC \ll 1$ the supercurrent $I_s(\phi) = I_c f(\phi)$ of the current essentially exceeds all other components and it is true equation for single junction interferometer (Ilichev et al. 2001a, b)

$$\phi = \phi_e - \ell f(\phi), \tag{1.47}$$

where ℓ is the normalized inductance of superconducting ring $\ell = \frac{2\pi L I_c}{\Phi_0}$. There is more precise method determination of CPR using radio frequency (RF) technique, which was proposed by Silver and Zimmerman (1967) and Rifkin and Deaver (1976) many years ago. Further development of this method presented by Ilichev et al. (2001), who showed that the CPR and the phase-dependent conductance could be extracted from experimental data. Results of measurements current through JJ as a function of the phase difference ϕ in symmetric 45° grain-boundary HTS junction presented in Fig. 1.16 (Ilichev et al. 1999). Recent development in fabrication of the JJs based on HTS achieved in Bauch et al. (2005). Measurements reveals that, YBCO based grain boundary tunnel junctions fabricated in Bauch et al. (2005) is highly hysteretic and Fig. 1.17 shows ratio of coefficients I_1 and I_2 determined by Fourier analysis of the CPR at various temperatures. Here I_1 and I_2 corresponds to the amplitudes of first and second harmonics. With decreasing T, modul of I_2 grows monotonically down to $T = 4.2$ K, while the I_1 component exhibits only a weak

Fig. 1.16 Experimental
CPR of symmetric $\pi/4$ grain
boundary junction on HTS

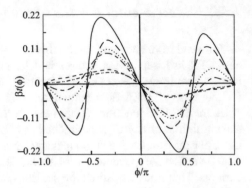

Fig. 1.17 Ratio of second
harmonic amplitude to basic
harmonic amplitude as
function temperature in
YBCO based grain boundary
junction

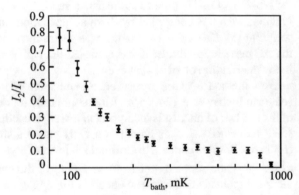

temperature dependence (Bauch et al. 2005). Furthermore, recently in an experimental study (Dirks et al. 2011) anharmonic CPR in graphene JJ was reported and corresponding theoretical calculations had been held in Black-Schaffer and Linder (2010).

Very recently topological insulators attached to superconductors attract great interest of researchers. The topological insulator offers a new state of matter topologically different from the conventional band insulator (Fu et al. 2007; Hasan and Kane 2010; Qi and Zhang 2011). When *SF* junctions are deposited on topological insulator, surface Dirac fermions acquire a domain wall structure of the mass. The CPR shows 4π periodicity, i.e., the shape of Josephson current has a fractional form $\sin(\phi/2)$ (Fu and Kane 2009; Ioselevich and Feigelman 2011). In study Williams et al. (2012) was reported JJ in hybrid superconductor-topological isolator devices, which reveals two peculiarities. $I_c R_N$ product for this structures inversely proportional with the width of the junction. Another moment related with a low characteristic magnetic field for suppressing supercurrent, i.e., Fraunhofer capture is different from traditional dependence $I_c(H)$ (Likharev 1986). The shape of CPR for S/TI structures presented in Fig. 1.18. Detail analysis of superconductor-topological insulator junctions is the subject of future investigations.

Fig. 1.18 CPR of
superconductor/topological
insulator junction

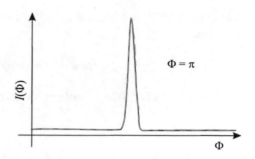

1.4 Dynamical Properties of JJs

1.4.1 Resistively Shunted Junction (RSJ) Model of JJs

In this section we discuss the IV curves of a tunnel JJs using an equivalent circuit
shown in Fig. 1.19. This circuit includes the effect of various dissipative processes
and the distributed capacity with so-called lumped circuit parameters (connection of
R and C in parallel). Using equivalent circuit of JJ we can write equation for current

$$C\frac{dV}{dt} + \frac{V}{R} + I_c \sin\phi + I_F = I, \tag{1.48}$$

where the first term is a displacement current, the second term is an Ohmic current,
the third term is a Josephson current, and $I_F(t)$ is the noise current source. Using
Josephson relation $\dot{\phi} = \frac{2e}{\hbar}V$, we get a second-order differential equation for the
phase ϕ

$$\frac{\hbar C}{2e}\frac{d^2\phi}{dt^2} + \frac{\hbar}{2eR_N}\frac{d\phi}{dt} + I_c \sin\phi + I_F = I. \tag{1.49}$$

In the last equation we take harmonic case of CPR, i.e., $\alpha = 0$ (anharmonic case
will considered in Sect. 1.5) and also we take the case of JJs with constant resistance,
i.e., $R(V) = R_N$, $R_s = 0$, $L_s = 0$. If introduce McCumber parameter $\beta_C = \frac{2\pi I_c R_N^2 C}{\Phi_0}$

Fig. 1.19 Circuit model of
RCLSJ in general case $f(\phi)$

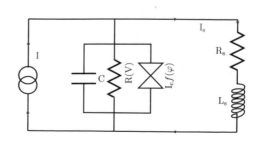

(McCumber 1968) and time units as $\frac{\Phi_0}{2\pi I_c R_N}$, we get JJ dynamic equation for harmonic CPR case

$$\beta_C \ddot{\phi} + \dot{\phi} + \sin\phi = i + i_F. \tag{1.50}$$

Derivative in Eq. (1.50) in respect to normalized time denoted by points above symbols.

1.4.1.1   IV Curve in the Case of $\beta_C \ll 1$ Low Capacitance Limit-Overdamped JJ

For the special case $\beta_C \ll 1$ (small capacitance limit), the above equation in the absence of fluctuations ($i_F = 0$) is reduced to

$$i = \dot{\phi} + \sin\phi, \tag{1.51}$$

which has an analytical solution with the period $T = \frac{2\pi}{\sqrt{i^2-1}}$

$$\phi = 2\arctan\left\{\frac{v}{i+1}\tan\frac{\omega_J t}{2}\right\} - \frac{\pi}{2}; \quad \omega_J = \frac{2e}{\hbar}V. \tag{1.52}$$

Averaging over period T leads to IV curve of following expression (Likharev 1986; Barone and Paterno 1982)

$$\langle v \rangle = \sqrt{i^2 - 1}. \tag{1.53}$$

IV curve of overdamped JJ using last expression presented in Fig. 1.20. It is clear that IV curve reveal nonhysteresis character. As shown by huge numbers of

Fig. 1.20   IV curve of overdamped JJ with harmonic CPR

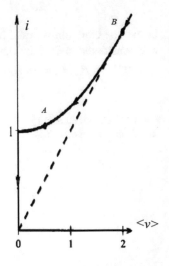

Fig. 1.21 Time dependent Josephson oscillations in different points of IV *curve*

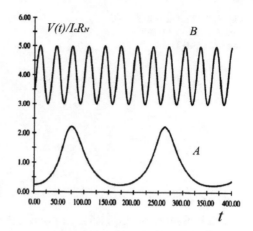

investigations, application of JJ with $\beta_C \ll 1$ in superconductivity electronics seems more convenient in contrast to one with $\beta_C \gg 1$ (see Chaps. 2 and 3). Figure 1.21 shows time dependent oscillation of phase and voltage in different points of IV curve of JJ with $\beta_C \ll 1$.

1.4.1.2 IV Curve in the Case of $\beta_C \gg 1$ High Capacitance Limit-Underdamped JJ

In this limit $\beta_C \gg 1$ (high capacitance limit) no analytical solution of Eq. (1.49). Numerical calculations shows that IV curve has two separate branch: S and R branch (Fig. 1.22). Figure 1.23 shows results of experimental measurements of IV curve of underdamped JJ. For $I > I_c$ there is only R-state of JJ. For $I < I_c$ there are R- and S-state of JJ. Important parameter of this IV curves is the return current I_R, at which arises switching of JJ from R-state to S-state. Calculation using simple resistive model of JJ junction I_R leads to result (Likharev 1986; Barone and Paterno 1982)

Fig. 1.22 Calculated IV curve of underdamped JJ with harmonic CPR

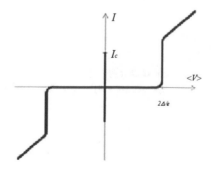

Fig. 1.23 Experimentally
measured IV *curve* of
underdamped JJ

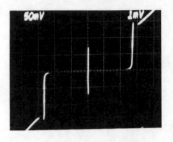

$$\frac{I_R}{I_c} = \frac{4}{\pi \sqrt{\beta_C}}. \tag{1.54}$$

Result presented by Eq. (1.54) in good agreement with experimental results and
numerical calculations (Likharev 1986; Barone and Paterno 1982).

Due to high intrinsic capacitance of tunnel JJ, a current-biased junction latches in
the resistive state after its $S \rightarrow R$ switching. As result bias current to be turned off
to perform the junction reset, i.e., $R \rightarrow S$ switching. This leads to lowering of clock
frequency of the latching logic circuit, which is about 1 GHz (Likharev 2012). For
this reason in JJ based logic circuits usually used junctions with $\beta_C \ll 1$ or shunted
tunnel JJ (Likharev 2012).

1.4.1.3   Shapiro Steps

In this subsection we consider the influence of harmonic external voltage $V = V_0 + v \cos \Omega t$ on IV curve of JJ. Corresponding equation for JJ phase has a form
(Shapiro 1963)

$$\phi(t) = \frac{2e}{\hbar}(V_0 t + \frac{v}{\Omega} \sin \Omega t + \phi_0). \tag{1.55}$$

Using expressions from handbook (Abramovitz and Stegun 1972)

$$\cos(a \sin \Omega t) = \sum J_{2n}(a) \exp(in\Omega t) \tag{1.56}$$

$$\sin(a \sin \Omega t) = \sum J_{2n+1}(a) \exp(in\Omega t) \tag{1.57}$$

and after averaging JJ current, it is clear that time-independent terms of current has
a peaks at voltage values $2 eV_0 = n\hbar\Omega$

$$\frac{I}{I_c} = (-1)^n J_n \left(\frac{2eV}{\hbar\Omega}\right) \sin \phi_0, \tag{1.58}$$

where $J_n(x)$ is Bessel function. This result experimentally observed in study (Shapiro
1963) and called as Shapiro steps (Fig. 1.24). It physically means that energy of

Fig. 1.24 Shapiro steps in
IV *curve* of JJ with harmonic
CPR

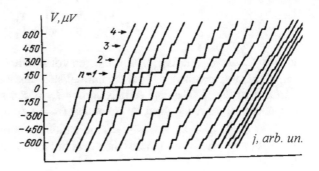

Cooper pair $2eV_0$ radiate as phonons with energy $\hbar\Omega$ or absorption of phonons leads to peaks on IV curve.

1.5 Influence of Anharmonic Effects of CPR on JJ Dynamics

1.5.1 Anharmonic Effects of CPR on IV Curve

For a long period, the real shape of the CPR was not considered as an important factor on dynamical properties of JJ. Tunnel JJ- SIS structures reveals $I_J(\phi) = I_c \sin \phi$, which executed experimentally with high quality for such junctions (see Sect. 1.3). The shape of the CPR, or more explicitly the energy and phase dependencies of spectral current density, become important in an analysis of the dynamic properties of JJ circuits. Small deviation from harmonic case $I_J(\phi) = I_c \sin \phi$ does not essentially affect the response of the junctions on a steady magnetic field and may be taken into account in the circuit design as an additional intrinsic inductance (see Zubkov et al. 1981), which has to be added to the geometrical one in the circuit simulation. In this section, we study IV curve of JJ with generalized CPR using relation

$$I_J = I_c(\sin \phi - \alpha \sin 2\phi).$$

The resistively, capacitively, inductively shunted JJ (RCLSJ) circuit shown in Fig. 1.19 is shunted by a small external resistor such that $R_s \ll R_N$ is satisfied (Likharev 1986; Whan et al. 1995; Cawthorne et al. 1998). Here R_N and R_s refer to normal state and shunt resistances respectively. As shown in the figure the tunnel JJ is replaced by three parallel current channels (Fig. 1.19). The total current across JJ is represented as a sum of supercurrent $I_J(\phi)$, the displacement current is represented by $I_D = C\frac{dV}{dt}$, the normal current due to the quasiparticles $I_N = \frac{V}{R(V)}$, where the voltage dependent junction resistance (Likharev 1986; Whan et al. 1995) is assumed to be

$$R(V) = \begin{cases} R_N & \text{if } |V| > V_g \\ R_{sg} & \text{if } |V| \le V_g \end{cases}, \tag{1.59}$$

given by $V_g = 2\Delta/e$ in Eq. (1.59) is the gap voltage that depends on the energy gap (i.e., Δ) of superconductor, R_N is the normal state resistance and R_{sg} is the sub-gap resistance of the JJ. The applied bias current I_{dc} is carried by the sum of the listed components (I_J, I_N, I_s) given by equations $I_J = I_c(\sin\phi - \alpha\sin 2\phi)$, $I_N = \frac{V}{R(V)}$ and $I_D = C\frac{dV}{dt}$: $I_D + I_N + I_J + I_s = I$, in which I_s denotes the current in the shunt branch. For simplicity, we ignored the effect of thermal noise throughout the section ($I_F = 0$).

Therefore, equations that corresponds to the circuit in Fig. 1.19 in general case in dimensionless form is

$$\dot{\phi} = v, \tag{1.60}$$

$$\beta_C \dot{v} + g(v)v + f(\phi) + i_s = i, \tag{1.61}$$

$$\beta_L \dot{i_s} + i_s = v, \tag{1.62}$$

where $\beta_C = (2e/\hbar)I_c C R_s^2$ is the McCumber parameter; $g(v) = R_s/R(v)$ is the normalized tunnel junction conductance; $i_s = I_s/I_c$ is dimensionless shunt current; $i = I/I_c$ is dimensionless external DC bias current; $\beta_L = (2e/\hbar)I_c L_s$ is the dimensionless inductance; $\tau = \omega_c t$ is the normalized time $\omega_c = (2e/\hbar)V_c$ is the characteristic frequency and $V_c = I_c R_s$ is the characteristic voltage. The relationship between β_C and ω_c can be given by $\beta_C = (\omega_c/\Omega_p(0))^2$ with the help of plasma frequency $\Omega_p(0) = \sqrt{2eI_c/\hbar C}$.

The solutions of Eqs. (1.60)–(1.62) are numerically obtained by using Matlab routine based on adaptive Runge–Kutta method (Canturk and Askerzade 2012). The time averaged voltage for the determination of I-V curve can be evaluated by using the expression:

$$\langle v \rangle = \langle \dot{\phi} \rangle = \frac{1}{\tau_{\text{rng}}} \int_0^{\tau_{\text{rng}}} v(\tau)d\tau, \tag{1.63}$$

where τ_{rng} is the sampling range. Notice that τ_{rng} in Eq. (1.63) is taken much longer than period of Josephson oscillations as well as relaxation oscillations. For that reason, the time averaged voltage in Eq. (1.63) is sometimes called as long-time average voltage in the literature.

In order to study the influence of second harmonic on the dynamics of the JJ, we, firstly, evaluate the critical current related to the amplitude of both harmonics. In this way, the normalized-critical current can be found as an extremum of the function $f(\phi) = \sin\phi - \alpha\sin 2\phi$

$$I_c/I_{c_0} = \max(f(\phi)), \tag{1.64}$$

where I_{c_0} is the critical current at $\alpha = 0$. The normalized critical currents with respect to anharmonicity parameter α is plotted in Fig. 1.25. As shown in the figure, I_c is

Fig. 1.25 Dependence of normalized critical current on anharmonicity parameter α

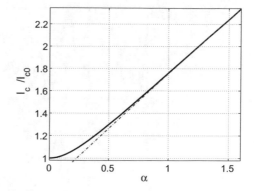

non-linear for small α, whereas, it is linear for large α values. Moreover, the linear dependence of critical current I_c was experimentally observed at large α in YBCO based JJs (Bauch et al. 2005). In addition, similar plot in Fig. 1.25 was obtained with an analytical expression for critical current given in Goldobin et al. (2007).

Two type dynamics of RCLSJ circuit presented in Fig. 1.26 can be explained using load-line analysis associated with an IV curve of JJ. The first case, shown in Fig. 1.26a, corresponds to relaxation oscillations in the circuit with parameters $i = 1.1$, $\beta_{C_0} = 1.11$, $\beta_{L_0} = 21.7$ at $\alpha = 0$. The relaxation oscillations in Josephson circuits, have been studied by many authors (Whan et al. 1995; Calander et al. 1981). Similar relaxation generator was used to study the dynamical properties of tunnel JJ comparator in Askerzade and Kornev (1994b). Second regime corresponds to the regular AC Josephson oscillations (Whan et al. 1995) and it is shown in Fig. 1.26b with parameters $i = 1.1$, $\beta_{C_0} = 2.22$, $\beta_{L_0} = 43.4$ at $\alpha = 0$. For non-zero anharmonicity parameter such as $\alpha = 0.4$ (see Fig. 1.26c), the amplitude of the Josephson oscillations becomes smaller, which is related to the effective capacitive properties of JJ. Such situation can be explained by increasing the critical current of JJ with anharmonic CPR (the detailed discussion is given below).

In study Canturk and Askerzade (2012) were performed numerical analysis of IV curve for different α values as shown in Fig. 1.27. In this figure, we have plotted IV curve of the system at different β_{C_0} and β_{L_0} values by using the same R_s, R_{sg}, and R_N from Table 1.1, similar (Whan et al. 1995). Notice that β_{C_0} and β_{L_0} refer harmonic case of the CPR (i.e., $\alpha = 0$). Similar results was obtained in Askerzade (2003) for only $\alpha = 0.2$ case due to the limited nature of the analytical calculations. Furthermore, it is difficult to study the details of the dynamics directly neither experimentally nor analytically, therefore, we can rely on numerical solutions of Eqs. (1.60)–(1.62) to study the influence of anharmonicity parameter α. As can be seen from Fig. 1.27a–c, the width of the hysteresis in the IV curve becomes larger by an increase in anharmonicity parameter α. Consequently, the presence of anharmonic CPR impacts an undesirable effect on inertial properties of the JJ. In addition, we repeat a few simulations by changing the sign of the anharmonicity parameter α as opposite to our calculations presented here. We observed that the hysteresis of the IV curve decreases

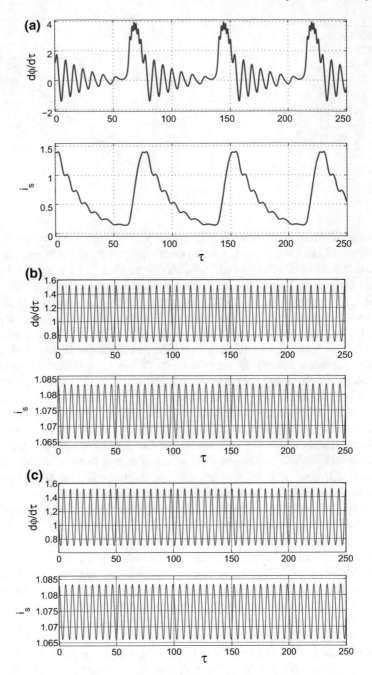

Fig. 1.26 Time dependence of dynamical variables: voltage $d\phi/d\tau$ and current through shunt branch i_s. **a** Computed at $i = 1.1$, $\beta_{C0} = 1.11$, $\beta_{L0} = 21.7$, and $\alpha = 0$. **b** Computed at $i = 1.1$, $\beta_{C0} = 2.22$, $\beta_{L0} = 43.4$, and $\alpha = 0$. **c** Computed at $i = 1.336$, $\beta_{C0} = 2.696$, $\beta_{L0} = 52.701$, and $\alpha = 0.4$

Table 1.1 Experimental parameters for JJ (Whan et al. 1995)

$T(K)$	$I_{c0}(mA)$	$V_g(mV)$	$R_{sg}(\Omega)$	$R_N(\Omega)$	$R_s(\Omega)$
4.22	0.550	2.91	50	3	1.1
7.60	0.275	2.09	15	3	1.1

Fig. 1.27 IV *curves* for various α values at $\beta_{C0} = 10$. **a** at $\beta_{L0} = 1$. **b** at $\beta_{L0} = 10$. **c** at $\beta_{L0} = 30$

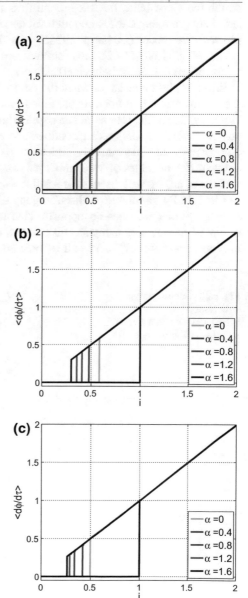

compared to its counter part in the presented plots. In general case, the sign of α is determined by the physical properties of barrier layer in Josephson structure (Buzdin 2005; Goldobin et al. 2007).

The size of hysteresis in IV curve is characterized by the return current I_R, at which the JJ switches from R-branch to S-branch in the IV curve. The relationship between return current and high values of McCumber parameter β_C using simple resistive model (Likharev 1986) gives the expression (1.54). If we take into account Eqs. (1.60)–(1.62), the return current versus McCumber parameter qualitatively reveals the similar behavior (Likharev 1986). The deviation from the expression in Eq. (1.54) becomes larger when the ratio of R_{sg}/R_N increases. On the other hand, the return current I_R is not only the function of β_C but also the function of α and β_L. However, it is difficult to obtain an explicit analytical expression for it. For that reason, numerical simulations are conducted to analyze the influence of anharmonicity parameter α and dimensionless inductance β_L on the normalized return current I_R/I_c at two different values of McCumber parameter (e.g., $\beta_{C_0} = 5$ and $\beta_{C_0} = 10$).

First of all, we will discuss the relationship between I_R/I_c and α which is shown in Fig. 1.28 for various β_{L_0} values. In general, the influence of the capacitance and inductance of the JJ has an opposite character on the junction impedance. That is one of the reactive element will damp the influence of the other. This implies that corresponding IV curve and associated hysteresis will be determined by the

Fig. 1.28 Return current with respect to anharmonicity parameter α. **a** at $\beta_{C0} = 5$. **b** at $\beta_{C0} = 10$

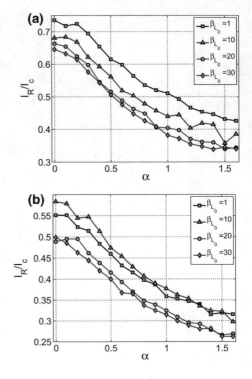

resulting impedance. At fixed β_{L_0}, by increasing α, the amplitude of the Josephson inductance $L_c = (\hbar/2e)/I_c$ decreases. As a result, the influence of the junction capacitance at non-zero α in the IV curve becomes dominant compared to harmonic case. If we compare the plots here with the results in study (Askerzade 2003), the normalized return current I_R/I_c, is accurately defined here in contrast to the result in Askerzade (2003). The reason for that, the author in Askerzade (2003) introduced an approximate solution of Josephson dynamics equation.

The exact value of return current I_R is sensitive to the characteristics of the junction in the subgap region. Usually, switching from R-state to S-state leads to an exponential decay of the voltage transient waveform. As mentioned in Likharev (1986), the voltage transient from R-state to S-state is accompanied by the slowly damping plasma oscillation. For that reason, we observed inaccuracy at some points on the curves presented in Fig. 1.28. The accuracy of this calculation for return current in Fig. 1.28a, b were roughly estimated by %5.

Besides, the relationship between the return current I_R/I_c with respect to dimensionless inductance β_L is illustrated in Fig. 1.29 for various α's. The junction circuit shown in Fig. 1.19 is shunted by a circuit of the serially connected inductance L_s and shunt resistor R_s. This means that the junction is shunted by impedance

Fig. 1.29 Return current with respect to dimensionless inductance β_{L_0}. **a** at $\beta_{C0} = 5$. **b** at $\beta_{C0} = 10$

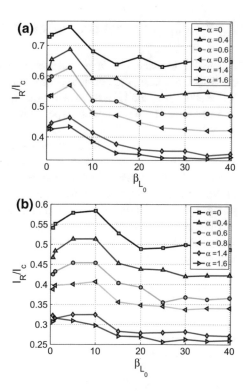

$$\frac{Z_s}{R_s} = \sqrt{1 + \left(\frac{\omega L_s}{R_s}\right)^2}. \tag{1.65}$$

At vanishing shunt inductance (i.e., $\omega L_s \ll R_s \ll R_{sg}$), we come to standard RSJ model with hysteresis on IV curve controlled by McCumber parameter $\beta_C = (2e/\hbar)I_c CR_s^2$. At an increase of the inductance L_s the impedance Z_s increases. On the other hand, the shunting effect in the high shunting inductance limit (i.e., $\omega L_s \gg R_s$) the impedance Z_s approaches to ωL_s. In this case, impedance Z_s becomes much greater than R_{sg} and R_N (i.e., $Z_s \gg R_{sg} \gg R_N$). As a result, McCumber parameter β_C can be determined by sub-gap resistance R_{sg}: $\beta_C(R_{sg}) = (2e/\hbar)I_c CR_{sg}^2$. Due to that reason, $\beta_C(R_{sg})$ will become greater than β_C. The return current I_R/I_c approximately reaches to constant values and lower than corresponding value for $\omega L_s \ll R_s$. As shown in Fig. 1.29, the crossover from one regime to another reveals peak character. Furthermore, as shown in Fig. 1.29a, b, the width of such peaks becomes wider for large McCumber parameter β_C. The inaccurate behavior similar to that in Fig. 1.29 is also observed on the curves in Fig. 1.28 due to the transient plasma oscillation in R→S switching.

1.5.2   Plasma Frequency of JJ Systems

It is useful to discuss influence of anharmonicity of CPR on small perturbations of the S state of JJ, i.e., the possible phase motion in the vicinity of equilibrium state ϕ_0. It is well known, the JJ dynamics has much in common with the motion of a particle in a potential of the washboard type Likharev (1986),

$$U(\phi) = -E_J \left(\cos\phi + i\phi - 1\right), \tag{1.66}$$

where i is the DC current expressed in the I_c (critical current) units, ϕ is the Josephson phase, and $E_J = \frac{\hbar I_c}{2e}$ is Josephson energy. In the case when the capacitance of the junction is sufficiently large, the junction may exhibit slowly decaying oscillations of the plasma phase at the bottom of the potential well (1.66). The frequency of these oscillations (plasma frequency) depends on the DC current and is given by the expression (see, e.g., Likharev 1986)

$$\Omega_p = \left(\frac{2eI_c}{\hbar C}\right)^{1/2} (1 - i^2)^{1/4}. \tag{1.67}$$

Relation (1.67) is usually observed with high precision in the JJs connected to a DC voltage source. In the case of $I = I_c \sin\phi$, the theory exhibits perfect agreement with experiment (Dahm et al. 1968). However, there are deviations from the behavior predicted by Eq. (1.67) in the JJs with anharmonic CPR.

1.5.2.1 Influence of Anharmonic CPR

The result presented by Eq. 1.67 is confirmed for small amplitudes of the AC current component. However, as the AC current amplitude grows, the agreement of Eq. (1.67) with the experimental values measured in deteriorates, which can be related to an anharmonic character of the potential $U(\phi) = -E_j(\cos\phi - \alpha\cos 2\phi + i\phi - 1)$ at large AC current amplitudes. The theoretical investigation of the effect of alternating current on the plasma frequency of the tunnel JJ simultaneously carrying DC and AC currents presented in Askerzade (2005b). The dynamics of a JJ can be described using the following equation (in this equation we use time units $\tau = \Omega_p(0)t$)

$$\ddot{\phi} + \frac{1}{\sqrt{\beta_C}}\dot{\phi} + \sin\phi - \alpha\sin 2\phi = i + i_d\sin\omega_d t. \tag{1.68}$$

For the further calculations we will use $\phi = \phi_0 + \phi_1$, where ϕ_0 is the equilibrium value and small deviation ϕ_1 obeyed the equation

$$\dot{\phi}_1 + \alpha\phi_1 = i + i_d\sin\omega_d t. \tag{1.69}$$

Using above presented mathematical expressions (1.56) and (1.57) from handbook (Abramovitz and Stegun 1972), we obtain for following expression for plasma frequency of JJ with anharmonic CPR

$$\frac{\Omega_p^2(a)}{\Omega_p^2(0)} = \cos\phi_0 + 2\alpha J_0(a)\cos 2\phi_0. \tag{1.70}$$

In last expression $J_0(a)$ is the Bessel function of zeros order, $\Omega_p(0) = \sqrt{\frac{2eI_c}{\hbar C}}$, $a = \frac{i_d}{\alpha^2+\omega_d^2}$. Equilibrium value ϕ_0 is determined from relation

$$\frac{i_0}{J_0(a)} = \sin\phi_0 + \alpha\sin 2\phi_0. \tag{1.71}$$

According to Eq. (1.70), an increase in the AC current component i_d leads to a decrease in the plasma frequency Ω_p. Thus, the presence of the AC component leads to renormalization of the plasma frequency (1.67) of the tunnel JJ.

The results of calculations using Eqs. (1.70) and (1.71) are presented in the Fig. 1.30 by the solid and dashed curves, respectively, in comparison to the experimental data (black circles) taken from Gronbech-Jensen et al. (2004). Result of calculations $\frac{\Omega_p^2(\alpha)}{\Omega_p^2(0)}$ as function of anharmonicity parameter α presented in Fig. 1.30. Nonsymmetric character of $\Omega_p^2(\alpha)$ is clear from calculations for different values of a (Fig. 1.31). There is minimum of $\frac{\Omega_p^2(\alpha)}{\Omega_p^2(0)} = 0.794$ at negative $\alpha = 0.3$ (Askerzade 2005b). At positive anharmonicity parameter α plasma frequency $\Omega_p^2(\alpha)$ decreased

Fig. 1.30 Plasma frequency
of JJ as function of AC
current amplitude

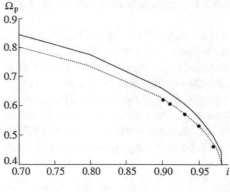

Fig. 1.31 Plasma frequency
of JJ as function of
anharmonicity parameter α

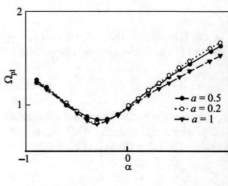

Fig. 1.32 Plasma frequency
of JJ as function of
amplitude of external AC
current for different
anharmonicity parameter α

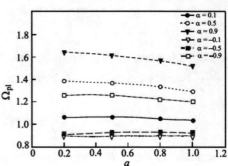

with increasing amplitude of oscillating part i_d. At negative anharmonicity parameter influence of α on the plasma frequency $\Omega_p^2(\alpha)$ is very small. In Fig. 1.32 presented results of calculations $\Omega_p^2(a)$ at different anharmonicity parameter α. At small α plasma frequency $\Omega_p^2(a)$ changes inconsiderable, while at high α influence of anharmonicity is important. At negative α influence of anharmonicity of CPR on plasma frequency is decreased.

1.5.2.2 Plasma Frequency of Layered Superconductors as Stack of Coupled JJs

Weak Josephson coupling between Cu–O planes in HTS can manifest itself in steps on the IV curves along the c axis (Kleiner and Muller 1994). As shown in studies (Mishonov 1991; Tachiki et al. 1994; Bulaevskii et al. 1995) a weakly damped Josephson plasma mode with the frequency Ω_p smaller than the superconducting gap Δ. The plasma mode gives rise to a narrow peak in the real part of the complex resistance $R(\omega)$ at low temperatures T. It is well known that at this temperature the quasiparticle damping is suppressed. A sharp peak in $ReR(\omega, H, T)$ has also been observed on $Bi - 2212$ single crystals in a DC magnetic field H (Kadowaki et al. 1997). The equations for $\phi(t)$ for a stack of coupled JJs have the form (Koyama and Tachiki 1996)

$$C_n \frac{dV_n}{dt} + G_n V_n + I_{cn} \sin \phi_n = I(t), \tag{1.72}$$

$$V_n - \alpha_0 (V_{n+1} + V_{n-1} - 2V_n) = \frac{\Phi_0}{2\pi} \frac{d\phi_n}{dt}, \tag{1.73}$$

where $I(t)$ is the external current density applied perpendicular to the layers, I_{cn} is the tunneling Josephson current density, V_n is the local voltage between the $n + 1$ and n th layers of thickness s spaced by d, α_0 is the charge coupling parameter, C is the specific interlayer capacitance, G is the interlayer conductance per unit area. Calculation of the complex resistance $R(\omega)$ taking into account Eqs. (1.72) and (1.73) leads to result (Gurevich and Tachiki 1999)

$$\frac{R(\omega)}{R_0} = \frac{if}{g} \left\{ N + \frac{\eta}{g - \eta f (f - i\gamma)/\sqrt{1 + 4\alpha_0/g}} \right\}, \quad R_0 = \frac{\hbar \Omega_p}{2e\sqrt{I_c^2 - I^2}}. \tag{1.74}$$

The first term in the brackets corresponds to a uniform sample (N is the number of layer in stack), and the second term related with the contribution of the localized plasma mode near grain boundary. Results of calculations for $Bi-2212$ single crystals presented in Fig. 1.33 (Gurevich and Tachiki 1999). As followed from Fig. 1.33, the strong Coulomb coupling can make the amplitude of the satellite peak, quite noticeable. In Gurevich and Tachiki (1999) it was shown that, measuring the satellite plasma peak in $ReR(\omega)$ at $\omega_0 < \Omega_p$ enables one to extract the charge coupling parameter α_0 and the supercurrent density I_0 across planar defects parallel to the ab planes.

1.5.2.3 Shapiro Steps in IV Curve of JJs with Anharmonic CPR

When AC signal is applied to JJ, its IV curve shows a set of Shapiro steps resulting from phase locking of Josephson oscillations (Shapiro 1963) (see Sect. 1.4 for detail). Analytical description of the Shapiro step dependence on the signal amplitude was

Fig. 1.33 Influence of
plasma oscillations on
complex resistance in
layered SC Bi-2212
(Gurevich and Tachiki 1999)

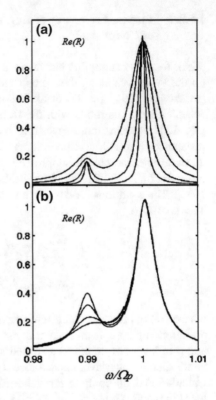

obtained only for a high-frequency limit in the frame of resistively shunted junction
(RSJ) model describing an overdamped junction with McCumber parameter $\beta_C \ll 1$
(Likharev 1986). In particular, a anharmonic CPR results in the generation of subharmonic Shapiro steps (Cassel et al. 2001), which may lead to instabilities in modes
of operation of Josephson devices. Results on analytical and computational investigations of high-frequency dynamics of JJs, characterized by non-zero capacitance
($\beta_C > 1$) and the second harmonic in the CPR are presented in Kornev et al. (2006).
Calculations gives result for step amplitude for harmonic case ($\alpha = 0$)

$$\Delta i_n = 2 \left| J_n \left(\frac{a}{\omega \sqrt{(\omega \beta_C)^2 + 1}} \right) \right|, \tag{1.75}$$

where a and ω is the amplitude and frequency of applied AC signal. The case of
$\beta_C = 0$ coincide with the well known RSJ model (Likharev 1986). In contrast to
harmonic case ($\alpha = 0$), there is subharmonic steps in IV curve and its amplitude is
given as (Kornev et al. 2006)

$$\Delta i_{(2n+1)/2} = 2\beta_C \left| \frac{J_n\left(\frac{a}{\omega\sqrt{(\omega\beta_C)^2+1}}\right) J_{n+1}\left(\frac{a}{\omega\sqrt{(\omega\beta_C)^2+1}}\right)}{\sqrt{(\omega\beta_C)^2/4+1}} \right|. \tag{1.76}$$

In case of JJ with anharmonic CPR ($\alpha \neq 0$) the following expressions for harmonic Shapiro step amplitudes are obtained (Kornev et al. 2006)

$$\Delta i_n = 2\max_\phi \left| J_n\left(\frac{a}{\omega\sqrt{(\omega\beta_C)^2+1}}\right)\sin\phi + \alpha J_{2n}\left(\frac{a}{\omega\sqrt{(\omega\beta_C)^2+1}}\right)\sin 2\phi \right|, \tag{1.77}$$

and subharmonic steps

$$\Delta i_{1/2} = 2\max_\phi \left| \sin\phi \left\{ \alpha J_1\left(\frac{2a}{\omega\sqrt{(\omega\beta_C)^2+1}}\right) + \beta_C \frac{J_1\left(\frac{a}{\omega\sqrt{(\omega\beta_C)^2+1}}\right) J_0\left(\frac{2a}{\omega\sqrt{(\omega\beta_C)^2+1}}\right)}{\sqrt{(\omega\beta_C)^2/4+1}} + \right. \right. \right.$$
$$\left. \left. \left. + 4\alpha^2\beta_C \frac{J_0\left(\frac{2a}{\omega\sqrt{(\omega\beta_C)^2+1}}\right) J_2\left(\frac{2a}{\omega\sqrt{(\omega\beta_C)^2+1}}\right)}{(\omega\beta_C)^2+1}\cos\phi \right\} \right| \right. \tag{1.78}$$

Figures 1.34 and 1.35 present the analytical results, as well as experimental data for both the c-oriented and c-tilted Nb/Au/YBCO junctions formed on NdGaO substrates (junction areas ranged from $10 \times 10\,\mu m^2$ to $30 \times 30\,\mu m^2$) (Komissinski et al. 2004; Wang et al. 2001). Similar results for subharmonic Shapiro steps was obtained in Kleiner et al. (1996) for c-axis YBCO/Pb tunnel junctions.

1.5.2.4 Influence of Anharmonic CPR on Dynamics of Long JJs

Flux dynamics in long JJ play important role in the modern superconductivity electronics. In terminology long JJ, the word "long" means that we take into dynamics equations the variation of the phase along one of the spatial coordinates (x). Furthermore, in contrast to Josepshon junctions with harmonic in CPR, there is no universal length scale at which changing of Josephson phase or at which the weak magnetic field is screened. Furthermore, the characteristic scales of x variation magnetic field screening can be different depending on the state. We also consider that the junction is short in the y direction in all states. The static phase distributions in the long JJs with anharmonic CPR carried out in Goldobin et al. (2007); Atanasova et al. (2010c). This model is described by double sine-Gordon equation for phase distribution in the static regime.

$$-\phi'' + \sin\phi - \alpha\sin 2\phi = \gamma, \quad x \in (-l, l) \tag{1.79}$$

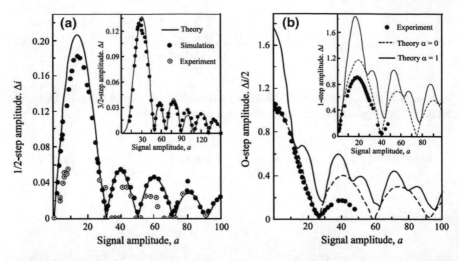

Fig. 1.34 *Left side* Dependences of the 1/2- and 3/2-step amplitudes on the applied signal amplitude a at frequency $\omega = 0.611$, $\beta = 35$ and $\alpha = 0$. *Solid line* corresponds to Eq. (1.76); *filled dots*, numerical simulation; and *empty dots*, experimental results for the *c*-oriented $Nb/Au/YBCO$ junctions. *Right side* Dependencies of the critical current amplitude $\Delta i/2$ (0-step) and the 1-step amplitude Δi (*in inset*) on the applied signal amplitude a at frequency $\omega = 1.62$ and $\beta = 4$. *Dashed* and *solid lines* correspond to Eq. (1.77) at $\alpha = 0$ and $\alpha = 1$ correspondingly, the *filled dots* correspond to experimental results for the c-tilted $Nb/Au/YBCO$ junctions (Wang et al. 2001)

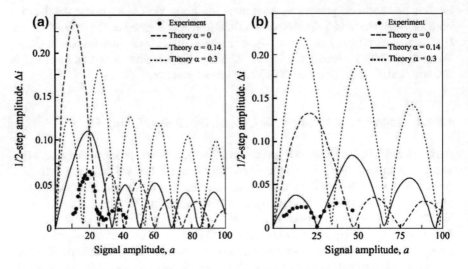

Fig. 1.35 Dependence of the 1/2-step amplitude Δi on the applied signal amplitude a at $\beta = 4$ for frequencies $\omega = 1.62$ (*left side*) and $\omega = 2.2$ (*right side*). *Dashed, solid* and *dotted lines* correspond to the step behavior given by Eq. (1.78) accordingly at $\alpha = 0$, $\alpha = 0.14$ and $\alpha = 0.3$. The *filled dots* are experimental data for the c-tilted $Nb/Au/YBCO$ junction (**b**) (Wang et al. 2001)

with the boundary conditions in the the following form (The notation $''$ on ϕ means second order spatial derivative)

$$\phi'(\pm l) = h_e. \tag{1.80}$$

In this equation γ is the external current in units of I_c, l is the semi-length of the junction, I_c and α is the anharmonicity parameters of CPR respectively. h_e is normalized external magnetic field. Stability analysis of $\psi(x, p)$ is based on numerical solution of the corresponding Sturm–Liouville problem

$$-\psi'' + q(x)\psi = \lambda\psi, \quad \phi'(\pm l) = 0, \tag{1.81}$$

with a potential $q(x) = \cos x - (\alpha/2)\cos 2x$. The minimal eigenvalue $\lambda_0(p) > 0$ corresponds the stable solution. In case $\lambda_0(p) < 0$ solution $\phi(x, p)$ is unstable. The case $\lambda_0(p) = 0$ indicates the bifurcation with respect to one of parameters $p = (l, \alpha, h_e, \gamma)$. Result of investigations in Atanasova et al. (2010a, b, c) show that anharmonicity in CPR significantly changes the shape and stability properties of trivial and phase distribution in long JJ.

In the harmonic case $\alpha = 0$ two trivial solutions $\phi = 0$ and $\phi = \pi$ (denoted by M_0 and M_π respectively) are known at $\gamma = 0$ and $h_e = 0$. Including of the second harmonic $\alpha \sin 2\phi$ leads to appearing two additional solutions $\phi = \pm \arccos(-1/2\alpha)$ (denoted as $M_{\pm ac}$). The corresponding λ_0 as functions of Eq. (1.79) coefficients have the form $\lambda_0[M_0] = 1 + 2\alpha$, $\lambda_0[M_\pi] = -1 + 2\alpha$ and $\lambda_0[M \pm ac] = \alpha(1 - (1/2\alpha)^2)$. The exponential stability of these constant solutions is determined by the signs of the parameters and by the α (Atanasova et al. 2010a, b, c) (Fig. 1.36). The full energy associated with the phase distribution $\phi(x)$ can be obtained from the following expression

$$F(p) = \int_{-l}^{l}\left[\frac{\phi'^2}{2} + 1 - q(x) - \gamma\phi\right]dx - h_e\Delta\phi. \tag{1.82}$$

Phase solution of Eq. (1.79) for harmonic case of CPR and in the case of $h_e = 0$ and $\gamma = 0$ at $l \to \infty$ has a form (Galperin and Filippov 1984)

$$\Phi_\infty^\pm = \phi(x) = 4\arctan(\exp(\pm x)) + 2\pi n \tag{1.83}$$

Fig. 1.36 Stability region as function of anharmonivcity parameter α

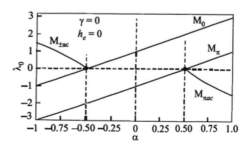

+ sign correspond to flux quantum, − sign correspond to antiflux quantum. At
small external fields h_e such distributions are flux quantum Φ^1, antiflux quantum
Φ^{-1} and their bound states $\Phi^1\Phi^{-1}$ and $\Phi^{-1}\Phi^1$. As external magnetic field h_e is
growing, more complicated stable flux quantum and bound states appear: $\Phi^{\pm n}$ and
$\Phi^{\pm nn}$ ($n = 1, 2, 3, \ldots$).

The energy of single flux quantum distribution Φ^1 limits to unit $F(\alpha \to 0) \to 1$
which corresponds to an energy of a single flux quantum Φ^1_∞ in a traditional "infinite"
junction model at $\alpha = 0$. With change of α the number of flux quantum

$$N(p) = \frac{1}{2\pi l} \int_{-l}^{l} \phi(x)dx \tag{1.84}$$

corresponding to the distribution Φ^1 is conserved, i.e., $\partial N/\partial \alpha = 0$. Here we have
$N[\Phi^1] = 1$. The results of influence of second harmonic CPR in $\lambda_0(h_e)$ calculated
in Atanasova et al. (2010a, b, c) and presented in Figs. 1.37 and 1.38. It is clear that
inclusion of anharmonicity affect in CPR significantly changes the shape and stability
properties phase distribution in long JJ.

Fig. 1.37 Dependence of
$\lambda_0(h_e)$ for Φ^1 at increase of
anharmonicity parameter α,
$2l = 10$, $\gamma = 0$

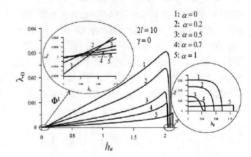

Fig. 1.38 Dependence of
$\lambda_0(h_e)$ for Φ^1 and Φ^{1*} at
increase of anharmonicity
parameter α, $2l = 10$, $\gamma = 0$

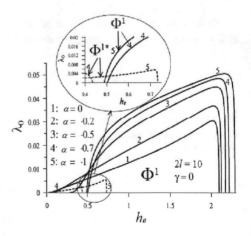

1.6 Macroscopic Quantum Dynamics of JJ

In terms of quantum-information science and searching for physical systems with macroscopic quantum behavior, superconductors have inherent advantages. It is generally accepted that, the detection of macroscopic quantum tunneling and energy-level quantization effects in conventional JJ circuits has enabled the fabrication of superconductive quantum devices based on LTS JJs. Superconductive quantum devices take advantage related with gap in the excitation spectrum, which favors less dissipation. On the other hand, the coherence of the superconducting state helps to achieve sufficiently long phase-coherence times. The interest in superconducting quantum systems is not only limited to quantum circuitry design, but also addresses fundamental issues on the nature of superconductivity and dissipation and the coherence in such objects. The result shows that the role of dissipation mechanisms in new superconducting systems has to be revised taking into account a macroscopic quantum degree of freedom in this device. In this section we present the observation of energy-level quantization in JJ, a clear signature of macroscopic quantum behavior. This indicate that the dissipation in a JJ is low enough to allow the formation of energy levels required for a qubit. We also present results of anisotropy and multiband effects on the JJ escape properties. At the end of this section effects of Coulomb blockade in small size JJs is presented.

1.6.1 Escape Rate in JJ at Low Temperatures

As is known, the JJ dynamics has much in common with the motion of a particle in a potential of the washboard type, given by Eq. (1.66). When current via JJ is ramped from $i = 0$ to $i < 1$, the junction is in the zero-voltage state in the absence of thermal and quantum fluctuations and the particle is confined to one of the energy minima (1.66), where it oscillates back and forth at the plasma frequency Ω_p. Due to thermal fluctuations, the junction may switch into a finite voltage state for a bias current $i < 1$ at finite temperature. This corresponds to the particle escaping from the well either by a thermally activated process (Kramers 1940) or by tunneling through the barrier potential, known as macroscopic quantum tunneling (Fig. 1.39) (Caldeira and Leggett 1983b; Devoret et al. 1985; Likharev 1983). In the regime of thermal

Fig. 1.39 Schematic description of escape in JJ in two different regimes (TA-thermal activation; MQT-macroscopic quantum tunneling)

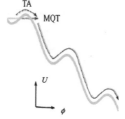

activation, the escape rate (the quality factor $Q = \Omega_p R_N C > 1$) is determined by Kramers (1940)

$$\Gamma_t(I) = a_t \frac{\Omega_p}{2\pi} e^{-\frac{\Delta U(I)}{k_B T}}, \tag{1.85}$$

where $\Delta U(I) = \frac{4\sqrt{2}}{3} E_J (1 - i)^{3/2}$ is the barrier height for i close to one and E_J is the Josephson energy. At sufficiently low temperature, the escape rate will be dominated by macroscopic quantum tunneling (Caldeira and Leggett 1983b; Devoret et al. 1985; Likharev 1983) for $Q > 1$ and i close to one it is approximated by the expression

$$\Gamma_t(I) = a_q \frac{\Omega_p}{2\pi} e^{-\frac{\Delta U(I)}{\hbar \omega_p} (1 + \frac{0.87}{Q})}, \tag{1.86}$$

where $a_q = \sqrt{\frac{864\pi \, \Delta U}{\hbar \omega_p}}$. In the derivation of the last expression, washboard potential was approximately replaced by cubic potential. The crossover temperature T_{cr} between the thermal and quantum regimes is given in Likharev (1983)

$$T_{cr} = \frac{\hbar \Omega_p}{2\pi k_B} \sqrt{1 + 1/4Q^2} - 1/2Q. \tag{1.87}$$

Crossover between two regimes, the switching histogram is schematically in Fig. 1.40. By lowering the temperature T the histograms move to higher currents and their width σ scales with the temperature down to T_{cr}. For $T < T_{cr}$, σ is independent of temperature. First experiments on macroscopic quantum tunneling in a JJ were carried out in study by Voss and Webb (1981) and Jackel et al. (1981). The behavior of the phase difference ϕ is deduced from measurements of the escape rate Γ of the junctions from its zero-voltage state. To determine the escape rate 10^4–10^5 events are typically collected for each set of parameters. The resulting switching current distribution $P(I)$ is used to compute the escape rate. The existence of quantized energy levels in the potential well of JJ was measured using spectroscopic method

Fig. 1.40 Schematic representation of crossover from thermal regime to quantum macroscopic tunneling

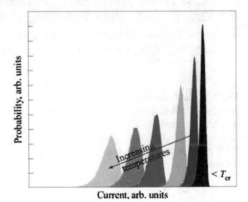

(Devoret et al. 1985; Martinis et al. 1985). The escape rate from the zero-voltage state was increased when the microwave frequency Ω corresponded to the energy difference between two adjacent energy levels. The quantum states can be observed using spectroscopic devices by inducing a resonant transition between the ground state and excited states by applying microwaves at frequencies $\omega_{0n} = (E_n - E_0)/\hbar$. The width of the first excited energy level is determined by the energy decay rate into the ground state, and is given by $1/\tau = Re(Y)/C = \omega_{01}/2\pi Q$ (τ is the lifetime of the first excited level). Here, $Re(Y)$ is the real part of the frequency dependent total shunting admittance Y related with dissipation, and Q is the quality factor of JJ. This set of experiments clearly indicated that ϕ is a quantum variable. Thermal energy must be sufficiently low to avoid incoherent mixing of eigenstates, and the macroscopic degree of freedom must be sufficiently decoupled from other degrees of freedom for the lifetime of the quantum states to be long on the characteristic time scale of the system (Caldeira and Leggett 1983b).

1.6.2 Influence of D-Wave Pairing Symmetry on Macroscopic Tunneling

In this subsection, we focus on how experiments on JJ macroscopic quantum behavior have been extended to novel types of structures and materials. Switching current distribution measurements has become a standard tool to investigate phase dynamics in systems with unconventional pairing and small sized structure. HTS are an example of unconventional systems, because of the d-wave order parameter symmetry (Massarotti et al. 2012). Quantum dynamics of a d-wave JJ was investigated in Bauch et al. (2006). In this study was fabricated highly hysteretic, tunnellike YBCO grain boundary JJs using the bi-epitaxial technique. Junctions were formed at the interface between a (103) YBCO film grown on a (110) SrTiO$_3$ (STO) substrate and a c-axis film deposited on a (110) CeO$_2$ seed layer. The orientation of the superconducting order parameter in the two electrodes is shown in Fig. 1.41. The enhancement of the

Fig. 1.41 The orientation of order parameter in d-wave JJ (Bauch et al. 2006)

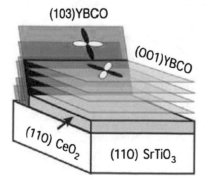

escape rate is measured under external microwave field to detect the energy levels. This allows us to analyze the bias current dependence of the energy-level separation and determine the width of the first excited level, which gives information on the dissipation processes in the junction. By repeatedly ramping the current I via JJ from zero at a constant rate, it was measured the switching current probability distribution $P(I)$. Microwaves at fixed frequency ω_{mw} were transmitted to the junction via a dipole antenna at a temperature below the crossover value, T_{cr}, separating the thermal and quantum regimes. When ω_{mw} of the incident radiation (or multiples of it) coincides with the bias current–dependent level separation of the junction, $\omega_{01}(I) = m\omega_{mw}$, the first excited state is populated. Here, m is an integer number corresponding to an m-photon transition from the ground state to the first excited state (Fig. 1.42). In these JJs the presence of a kinetic inductance and a stray capacitance determine the main difference in behavior (Rotoli et al. 2007). The YBCO based JJ is coupled to this LC-circuit and the potential becomes two-dimensional.

HTS may be an interesting reference system for novel ideas on key issues on coherence and dissipation in JJ systems. Low-energy quasiparticles have represented since the very beginning a strong argument against the occurrence of macroscopic quantum effects in these compounds. Quantum tunnelling regime of the phase leads to fluctuating voltage across JJs which excites the low-energy quasiparticles specific for d-wave junctions, causing decoherence. Experiments on $Bi_2Sr_2CaCu_2O_8$ stack JJs have been aimed to increase the crossover temperature T_{cr} and to clarify the nature of internal JJs. In these JJs the nodes of the d-wave superconducting order parameter are not expected to affect significantly macroscopic quantum tunneling. Josephson coupling between CuO_2 double layers has been proved, and most of the compounds behaved like stacks of SIS JJs with effective barriers (Critical current density J_c typically 103 A/cm^2) (Rotoli et al. 2007; Jin et al. 2006). However, at high-voltages caution is required when extracting information because of possible unavoidable heating problems. T_{cr} has been reported to be about 800 mK, remarkably higher than those usually found in LTS systems. By using spectroscopic method, the

Fig. 1.42 Energy levels in washboard potential of JJ $U(\phi)$

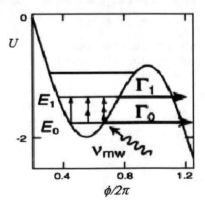

structure of intrinsic JJ stacks have been considered responsible for a remarkable enhancement of the escape rate (Jin et al. 2006).

1.6.3 Influence of Multigap Nature of Superconductivity on Escape Rate

The macroscopic quantum tunneling in JJ based on multigap superconductors was investigated in Ota et al. (2011). Such JJ can be fabricated by using recently discovered Fe-based superconductors or MgB_2 (Ota et al. 2009). These superconductors have multiband characters and their Fermi surfaces reveals disconnected structure (Askerzade 2012). Besides, the superconducting gap can be individually well defined on each Fermi surface. In JJ based on multigap superconductors one may expect that the superconducting tunneling current has multiple channels between the two superconducting electrodes (Yerin and Omelyanchouk 2010; Agterberg et al. 2002). In study Ota et al. (2009) constructed a theory for the quantum switching in JJs based on multiband superconductors. Supercurrent density between two superconducting electrodes is given by the sum of the currents in the two tunneling channels as $I = I_{c1} \sin \phi_1 + I_{c2} \sin \phi_2$, where I_{ci} is the Josephson critical currents in the ith tunneling channel (Fig. 1.43). Lagrangian formalism and instanton approximation (Ota et al. 2011) leads to macroscopic quantum tunneling escape rate for this case

$$\Gamma_t(I) = 12\omega_P(i)\sqrt{\frac{3V_0}{2\pi\hbar\omega_P(i)}}e^{-\frac{36V_0}{5\hbar\omega_P(i)}}, \qquad (1.88)$$

where $\omega_p(i) = \Omega_p((1-\varepsilon)^2 - i^2)^{1/4}$, $V_0 = \hbar^2\Omega_p^2(i)\cot^2(\frac{\phi_0}{3E_C})$, $(1-\varepsilon)\sin\phi_0 = i$. In the derivation of Eq. (1.88), zero point motion of the phase was considered. Results of calculations (Ota et al. 2011) presented in Fig. 1.44 shows a contour map of the ratio Γ/Γ_0 in the (i vs ω_P/ω_{JL}) plane with Γ_0 being the escape rate without correction, i.e., $\varepsilon = 0$. It is clear that the escape rate is enhanced in a wide parameter region. In particular, the enhancement is pronounced in the region of large Ω_P/Ω_{JL}. As shown in Eq. (1.88), the Josephson energy E_J is renormalized by the zero-point motion of phase and the renormalized one is decreased from the bare one because $\varepsilon > 0$

Fig. 1.43 Schematic view of Single-Band/ I /Two-Band JJ

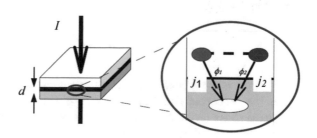

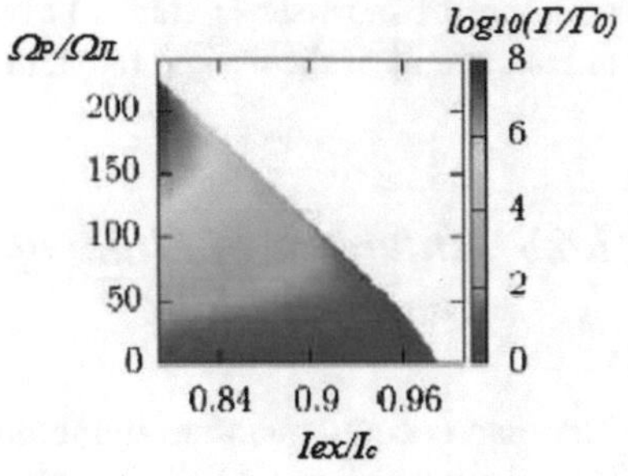

Fig. 1.44 Ratio of Γ/Γ_0 as function of bias current in Single-Band/ I /Two-Band JJ

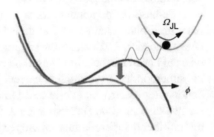

Fig. 1.45 Renormalization of washboard potential in Single-Band/ I /Two-Band JJ

(Fig. 1.45). In addition, it is clear that the zero-point fluctuation becomes larger as the frequency of the Josephson-Leggett mode decreases (Askerzade 2012). It is useful to note that with a lowering of Ω_{JL}, escape rate is enhanced considerably.

1.6.4   Effect of Coulomb Blockade in Nano-Size JJs

In recent years the study of JJs with small capacitance becomes very attractive. The interest in superconducting circuits with such JJs are related with quantum behavior of these macroscopic systems. For example, the small-capacitance JJs can transfer individual Cooper pairs and generate Bloch oscillations (Likharev and Zorin 1985; Nakamura et al. 1999). The small-capacitance JJs are characterized by the finite ratio E_J/E_C; the Josephson coupling energy E_J and the charging energy $E_C = 4e^2/2C$ (C is the junction capacitance). The important peculiarity of these JJs is their nonlinear differential capacitance C_B related to the local curvature of the zero Bloch energy band (Fig. 1.46) (Likharev and Zorin 1985; Zorin 2006)

$$C_B^{-1}(q) = \frac{d^2 E(q)}{dq^2}, \qquad (1.89)$$

where $E_0(q)$ is the ground state energy of the junction. E_0 periodically (the period is equal to 2e) depends on the quasicharge q which is a good variable in contrast to Josephson phase (see Fig. 1.46). Variable q is analog to the quasi-momentum of an electron moving in the periodic potential of a crystal lattice. The circuit consisting of

Fig. 1.46 Reverse Bloch capacitance $C_B^{-1}(q)$ as function of quasi-charge

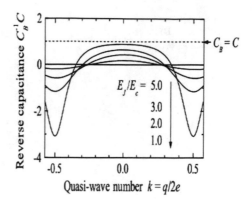

Reverse capacitance $C_B^{-1}C$

$E_J/E_c = 5.0$
3.0
2.0
1.0

$C_B = C$

Quasi-wave number $k = q/2e$

a small-capacitance JJ biased by the ideal source of the classical current I_t (Likharev and Zorin 1985) is described by the Hamiltonian $H = H_0 + H_I$:

$$H_0 = \frac{Q^2}{2C} - E_J \cos\phi; \quad H_I = -\frac{\Phi_0}{2\pi}\phi I(t). \qquad (1.90)$$

Here H_0 is the the junction Hamiltonian and H_I is the source part. The operator of the charge Q is conjugate to the phase operator ϕ and is equal to $Q = -i2e\frac{d}{d\phi}$. The quasiparticle tunneling is neglected for the sake of clarity. The eigenenergies $E_n(k)$ of the base Hamiltonial H_0 are periodic functions of the continuous quasi-wave number $k = q/2e$ or the quasicharge q, with the period equal to 1 and 2e, respectively; the band index $n = 0, 1, 2, \ldots$ (Likharev and Zorin 1985). Electrical diagram of the current-biased nano-sized JJ and the equivalent circuit of this junction is presented in Fig. 1.47.

Bloch inductance which was introduced the first time in the study (Zorin 2006) is given by

$$L_B = \frac{L_B^{\max}}{(1+\xi^2)^2}; \quad L_B^{\max} = \frac{\Phi_0}{2\pi I_c}; \quad |\xi| \le 1 \qquad (1.91)$$

Fig. 1.47 Circuit model of JJ with Coulomb blockade

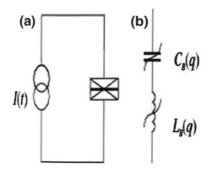

(a)

$I(t)$

(b)

$C_B(q)$

$L_B(q)$

Fig. 1.48 Quasicharge
dependence of Bloch
inductance $L_B(q)$

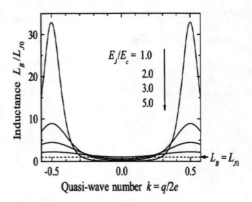

at $k = \frac{1+\kappa\xi}{2\,\mathrm{mod}\,1}$, where $\kappa = \frac{E_J}{E_C}$. It is clear that in contrast to the phase-dependent Josephson inductance $L_J(\phi)$, the Bloch inductance $L_B(k)$ is a positive periodic function of the quasicharge (see Fig. 1.48). Physically, L_B still characterizes the kinetic properties of supercurrent. Taking into account Bloch inductance, the equation of motion for quasicharge q takes the form (Zorin 2006):

$$L_B(q)\frac{d^2q}{dt^2} + R_N\frac{dq}{dt} + V(q) = V. \qquad (1.92)$$

Thus, for the describing its effect on the quasicharge dynamics in circuits with Bloch inductance and Bloch capacitance should be used. The effect should manifest itself by an effective renormalization of the Bloch capacitance in the vicinity of the degeneracy points of the Cooper-pair boxes and by a hysteretic behavior of the IV curves of long one-dimensional arrays. As noted in Zorin (2006), the Bloch inductance may be applied in the dispersive quantum nondemolition readout of Josephson qubits (see Chap. 4).

Chapter 2
Analog Superconductivity Electronics

Abstract In this chapter we present results of achievement of modern analog superconducting electronics. In first section we discuss physical foundation and characteristics of superconducting edge bolometers. The influence of HTS materials on the characteristics is analyzed. Second section deals with description of samplers based on JJs. Two types of comparators-on single junction and balanced comparators are discussed. Presented results of linear theory for the estimation of time resolution of JJ comparators. Comparators using RSFQ technique discussed and ultimate characteristics presented. Third section is devoted to AC and DC SQUID magnetometers. Firstly, single and two-junction interferometers, the influence of d-wave order parameter symmetry on the characteristics are described. Furthermore, using these interferometers as sensitive elements of SQUIDs and characteristics is presented. Next section devoted to description of current state of superconducting microwave devices. Last topics in this chapter are using JJs in meteorology and description of multi-terminal superconducting devices.

2.1 Superconducting Bolometers

Superconducting bolometers is the *simplest devices* based on superconducting materials. It is well known, that the physical principle operation of the edge-transition bolometers is based upon the steep drop in their resistance, R, at critical temperature T_c. The bolometer converts the high-frequency current into an output signal at the frequency of modulation. Superconducting antenna bolometers can be used with various antennas, such as bow-tie, log-periodic or log-spiral. The main advantage of a superconducting bolometer is that since it operates as thermal detector it may be sensitive over a very wide electromagnetic spectrum. Antenna bolometers are characterized by several performance parameters. Some of these parameters are *NEP* (noise equivalent power), responsivity r and response time. *NEP* is defined in the same way as for any thermal detector. For a frequency band of $1Hz$ it is given by Leonov and Khrebtov (1993)

© Springer International Publishing AG 2017
I. Askerzade et al., *Modern Aspects of Josephson Dynamics
and Superconductivity Electronics*, Mathematical Engineering,
DOI 10.1007/978-3-319-48433-4_2

$$(NEP)_{min} = \frac{P_b^2}{\varepsilon} + \frac{4kT^2G}{\varepsilon^2} + \frac{4kT^2R}{S^2} + \frac{V_{1/f}^2}{S^2} + \frac{V_{amp}^2}{S^2}. \tag{2.1}$$

Where k is the Boltzmann constant, T is the bolometer temperature, P_b is the fluctuation power of background radiation, R is the bolometer resistance, ε is the absorption coefficient and G is the effective thermal conductance characterizing the thermal coupling between the bolometer and the heat sink. The first term is (2.1) is related with background radiation, the second term originates from the random energy exchange between the bolometer and the heat sink via the thermal conductance G. The third term takes into account the Johnson noise of the bolometer resistance, the fourth is determined by the excess voltage $1/f$ noise $V_{1/f}$ of a superconducting film. The fifth term is related with the contribution of the amplifier noise V_{amp}. As shown by estimation, at nitrogenim temperature $T = 4K$ noise equivalent power NEP for LTS can be about $(NEP)_{min} = 10^{-14}\ Watt/(Hz)^{1/2}$ (Karmanenko et al. 2000).

Since the discovery of HTS, many works have been focused on the application of these materials in different types of bolometers for the near and far infrared wavelength regime (Fardmanesh and Askerzade 2003; Bozbey et al. 2003). The responsivity versus modulation frequency and versus the temperature of superconducting bolometers has previously been investigated and reported in Leonov and Khrebtov (1993). A simple RC-model for the amplitude of bolometric response results in the following expression (Fardmanesh 2001)

$$r = \frac{\varsigma I}{G(1 + j\omega\tau)}\frac{dR}{dT}, \tag{2.2}$$

where $\tau = C/G$, I is the DC bias current, ς is the fraction of the incident power absorbed by the bolometer, ω is the modulation frequency, r is the frequency-dependent responsivity in units of $V W^{-1}$, and G and C are the thermal conductance and the heat capacity of the bolometer, respectively. As shown in (2.2), thermal conductance G is a major parameter in the amplitude of the response function. According to (2.2) the temperature-dependent phase of the response can be calculated by the equation

$$\tan\theta = -\frac{\omega C}{G}. \tag{2.3}$$

It is important to understand the factors that determine G and C of the bolometers as a function of the temperature, frequency and other parameters of the bolometer configuration. In the study (Bozbey et al. 2003) developed thin-film YBCO on crystalline MgO bolometers. Below presented on the phase of the response to infrared signal of the edge-transition bolometers versus temperature at high and low modulation frequencies. In study (Karmanenko et al. 2000) it was proposed a model that can explain the observed discrepancy in the measured response versus temperature compared to that expected from other models. According to presented model, the thermal constants G and C change as a film goes into the superconducting state. This might also partly be accounted for some observed non-bolometric component of the response in these types of bolometers.

Fig. 2.1 Sample
configuration of YBCO
based bolometer (d_s is
substrate thickness)

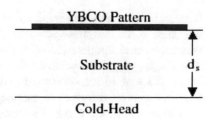

The typical configuration of the studied samples is shown in Fig. 2.1. A sample holder was designed and made of high purity and high conductive oxygen-free copper. The heater is made of resistive paste on a sapphire substrate using hybrid microelectronics technology and can control the temperature with a DC current up to maximum of about 200 mA with a precision of 0.1 K. As shown in Fig. 2.1, there are two thermal boundary resistances at the substrate interfaces, one at the film interface and the other at the holder interface. There are also two major bulk areas which constitute the overall heat capacity of the samples, one due to the superconducting film and the other due to the substrate material. The thickness of the superconducting YBCO film was about 200 nm. The phase and magnitude of the response versus temperature at different frequencies were measured using a lock-in amplifier. A light emitting diode with a peak radiation wavelength of about 0.85 μm was used as the radiation source for the response versus modulation frequency studies in all the measurements. In analyzing the different operation regimes of the edge-transition bolometer parameters, the frequency dependence of the thermal diffusion length plays an important role. It is well known that thermal diffusion length into substrate, L_f, is determined as Hu and Richards (1989), Hwang et al. (1979)

$$L_f = \left(\frac{D}{\pi f}\right)^{1/2},$$ (2.4)

where $D = K/C$ is the thermal diffusivity of the substrate material, and K and C are the thermal conductivity and specific heat (per unit volume) of the substrate material, respectively. At high chopping frequencies, L_f becomes comparable to or smaller than the substrate thickness changing G. Then G is limited by the thermal conductance of the substrate material and can be obtained from

$$G = \frac{KA}{L},$$ (2.5)

where L and A are the thickness of the substrate and the area of the superconducting pattern, respectively.

The phase of the response of the samples is found to be more sensitive to the values of the characteristic parameters of the bolometers than the magnitude of the response. The phase of the response has shown a strong variation as the sample goes through the normal–superconducting transition both at low and high modulation frequencies also showing modulation frequency dependence. At low frequencies, the phase of the response starts to decrease as the temperature decreases below about the onset temperature of the normal to superconducting transition (i.e., at $T < T_{c-onset}$). Opposite behaviour has been observed at high modulation frequencies, i.e. the phase of the response starts to decrease at $T < T_{c-onset}$. This clearly indicates that large changes in the phase and magnitude of response are mainly associated with the differences in the conduction mechanisms in the normal and superconducting states, rather than the well-known temperature dependence of the interface boundary resistance.

2.1.1   Low Frequency Limit

At low modulation frequencies the phase of the response versus temperature increases by decreasing the temperature starting at $T < T_{c-onset}$ as shown in Fig. 2.2. The respective amplitude of the response normalized to the measured dR/dT of the superconducting film at $T < T_{c-onset}$ for low modulation frequencies is shown in Fig. 2.3. The amplitude of the response in Fig. 2.3 increases above the expected value determined by the dR/dT of the film as the temperature is lowered towards the T_{c-zero}. Such behavior would be consistent with a decrease of the thermal conductance of the bolometer. The variation of the heat capacity at low temperatures, reveals typical T^3 dependence behavior (Barron et al. 1959). At low modulation frequencies ($\omega = 20\,Hz$), the thermal conductance in the studied samples is determined by the substrate/cold-head interface conductance, or the so called Kapitza boundary resistance, caused by the acoustic mismatch impedances of the interfaced materials (Fardmanesh 2001). According to (2.4), by lowering the frequency the thermal diffusivity length into the substrate becomes longer than the substrate thickness and phonons emitted by the superconducting film into the substrate will reach the substrate/cold-head boundary. The observed thermal boundary conductance at

Fig. 2.2 Phase of the response versus temperature at low modulation frequency 20 Hz and 680 μA

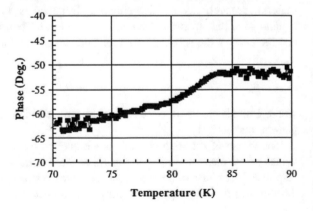

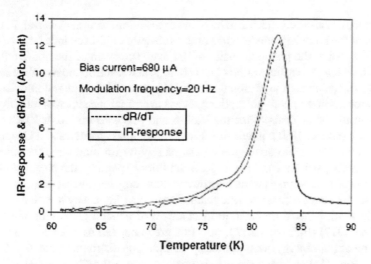

Fig. 2.3 Amplitude response versus temperature at modulation frequency 20 Hz and 680 μA

the substrate/cold-head interface can be explained by the theory of Kapitza conductance. It is well known that Kapitza conductance is the derivative of the net heat flux transmitted across an interface with respect to the temperature difference between the two materials (Swartz and Pohl 1989). Therefore, if we consider an interface between materials A and B, we can write G_K as

$$G_K = \frac{1}{V}\frac{\partial}{\partial T}\sum_k^A\sum_j \hbar\omega_{kj}n(\omega_{kj}, T)v_{kjz}t_{kj} = \frac{1}{V}\frac{\partial}{\partial T}\sum_k^B\sum_j \hbar\omega_{kj}n(\omega_{kj}, T)v_{kjz}t_{kj},$$

(2.6)

where the labels A and B on the sums indicate that all quantities in each sum correspond to phonons incident on the interface from the materials A and B sides, respectively. In Eq. (2.6), ω_{kj} is the phonon frequency, v_{kjz} is the component of the phonon group velocity normal to the interface, $n(\omega_{kj}, T)$ is the Bose–Einstein distribution function, t_{kj} is the probability that a phonon of wave vector k and polarization j will be transmitted across the interface between materials A and B, and V is the volume. The Kapitza conductance can be calculated at low temperatures ($\hbar\omega_D/kT \gg 1$) analytically taking into account Debye phonon density of states, which can be obtained from

$$D(\omega) = \frac{1}{V}\sum_{kj}\sum_j \delta(\omega - \omega_{kj}) = \frac{\omega^2}{2\pi^2 c^3},$$

(2.7)

where c is the velocity of sound in the material. As a result of calculations, the thermal conductance has T^3 characteristics at low temperatures and remains constant at high temperatures. The phonon system in each material in the above calculations is considered to be of the equilibrium nature. However, in the calculation of the boundary conductance of the substrate/cold-head interface we should take into account the phonon density of the state spectrum of the non-equilibrium phonons, which propagate into the MgO substrate and reach boundary substrate/cold-head (the diffusion regime of propagation of phonons). Hence, one should determine the differences in the photo-absorption process in normal and superconducting states. As shown by the analysis of thermalization and photoabsorption in YBCO superconductors (Bluzer 1991), interaction between hot quasi-particles and Cooper pairs condensate continues to divide the excess energy by forming three quasiparticles. The hot quasi-particles continue to break additional pairs by the electron–electron interaction process as they thermalize toward the energy gap by discrete energy steps. As a result we have a characteristic feature-a jump at frequency $\omega = 2\Delta(T)$ due to the activation of the recombination mechanism of phonon generation (Chang and Scalapino 1977) (for $\omega < 2\Delta(T)$, phonons are generated as a result of the energy relaxation of high energy quasi-particles). The phonon distribution function (product of $D(\omega)n(\omega, T)$) has fairly narrow maximum near $\omega = 2\Delta(T)$ (Sergeev and Reizer 1996), i.e. in our calculation we should replace Debye phonon density of states by the Einstein phonon density function

$$D(\omega) = \omega^2 \delta(\omega - 2\Delta(T)). \qquad (2.8)$$

Assuming that transmission probability and normal component of the group velocity to the interface to be independent of temperature, Eq. (2.6) can be rewritten as

$$G_K = \frac{1}{V} \sum_j^A \int_0^{\omega_{Debye}} d\omega \hbar \omega_{kj} D(\omega) \frac{dn(\omega_{kj}, T)}{dT} v_{kjz} t_{kj}, \qquad (2.9)$$

where for $dn(\omega, T)/dT$ we have the following expressions:

$$\frac{dn(\omega_{kj}, T)}{dT} = \left\{ \begin{array}{ll} \frac{k}{\hbar\omega}, & \frac{\hbar\omega}{kT} \ll 1 \\ \frac{k}{\hbar\omega}x^2 \exp(-x), & x = \frac{\hbar\omega}{kT} \gg 1 \end{array} \right\}. \qquad (2.10)$$

It is clear that at high phonon frequencies, $2\Delta(T)$ is about the equilibrium temperature of the substrate T and the contribution of the non-equilibrium phonons to the heat transfer has exponential behavior. As shown by the above calculations, at temperatures close to critical temperature $2\Delta(T) \ll T_c$, where $\Delta(T) = \Delta_0(1 - T/T_c)^{1/2}$ (standard BCS theory), low frequency phonons do not influence heat transfer process and boundary conductance at substrate/cold-head. Hence, the calculated normalized conductance based on expressions (2.9) and (2.10) are in good agreement with the experimental data for the thermal conductance versus temperature at low modulation frequency. The discrepancy between the experimental data and calculations at temperatures close to T_c shown in Fig. 2.4 is interpreted to be associated with the tendency of the emitting phonon frequency to zero. Here we also consider superconducting materials without magnetic flux quantum. It is clear that temperatures close

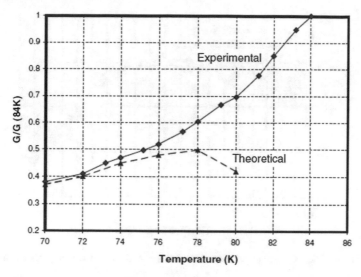

Fig. 2.4 Temperature dependence of normalized Kapitza resistance of bolometer

to the T_c can easily lead to Abrikosov vortexes in the films. In the equilibrium, flux quantums are present in the film and move due to the thermal activations. The optical photons also create additional magnetic flux quantum and the flux motion (creep and flow). This factor can lead to the non-bolometric component in the optical response, not considered in the above calculations. In our opinion, discrepancy between theory and experiment can also be related to flux dynamics near T_c. This argument is suggested by the experimental data from Fardmanesh (2001). As shown in this paper, for the samples with sharper transition, the change in the phase of the response and the discrepancies at low frequencies are smaller. One of the possible reasons for the broadening of the transition into a superconducting state is the penetration of flux quantum into the samples. Another possible improvement of presented model is connected with using realistic behavior of the order parameter close to T_c and d-wave characteristics of the pairing in YBCO compounds. On the other hand, influence of the order parameter relaxation near T_c on the bolometer characteristic should also be taken into account for more accurate calculations near critical temperature.

2.1.2   High Frequency Limit

The variation of the phase of the response versus temperature at high modulation frequency (e.g., $\omega = 10\,\text{kHz}$) is shown in Fig. 2.5. At frequencies above the knee frequency, f_L, the phase of the response decreases as the temperature is lowered. To describe the heat transfer in this regime we imply the phonon radiation limit (Bluzer 1991) (ballistic regime of propagation of phonons). In the high-frequency

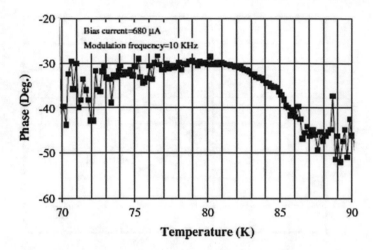

Fig. 2.5 Phase of the response versus temperature at high modulation frequency 10 kHz and 680 μA

regime, all phonons from the superconducting film are assumed to transmit across the film–substrate boundary and absorbed in the substrate media. That is, in Eq. (2.6), transmission probability t_{kj} can be considered 1. Such approximation is valid due to the low Kapitza resistance of the film–substrate boundary (Hu and Richards 1989). As shown by calculations (Richardson et al. 1992), the thermal conductivity in HTS films has different behavior in contrast to conventional LTS. In cuprate compounds, thermal conductivity below T_c versus temperature has a peak near the T_c which is lowered by decreasing the film thickness (Richardson et al. 1992). The thermal conductance of the studied devices at high modulation frequencies is previously shown to be mainly governed by thermal conductivity of the substrate material taking into account the processes in superconducting film (Hu and Richards 1989). The thermal conductivity of the substrate can also increase as the temperature decreases below the $T_{c-onset}$. This is due to enhancement of the mean-free path of phonons emitted from the superconducting film into the substrate. As a result, the phase of the response in the temperature interval 75 K $< T <$ 85 K decreases and at lower temperatures, $T <$ 75 K, increases. The corresponding amplitude of the response at high frequencies is shown in Fig. 2.6. Response at temperatures $T <$ 75 K can be closer to the dR/dT curve. The discrepancy in amplitude of response in the temperature interval 75 K $< T <$ 85 K is interpreted to be related to the variation of the thermal conductivity in the substrate material.

2.2 Superconducting Samplers

The considerable nonlinearity of the IV curve of JJs makes them promising for digital and pulsed devices (Hamilton et al. 1981; Hayakawa 1983). Super-high-speed switches have been built around complex distributed JJs and interferometers, where

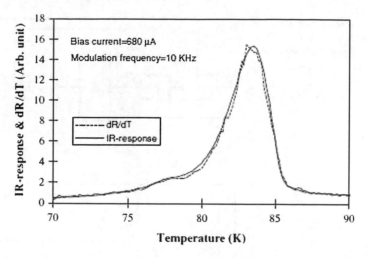

Fig. 2.6 Amplitude response versus temperature at modulation frequency 10 kHz and 680 μA

switching is induced by suppressing the critical current of the system (Zappe 1975; Fulton et al. 1977). Switching cells on JJs can be viewed as building blocks for samplers. These devices make it possible to display the waveform of weak ultrashort (picosecond) pulses (Wolf et al. 1985a; Akoh et al. 1983). The schematic of a comparator sampler is shown in Fig. 2.7a; the signal waveform and the operation of the device, in Fig. 2.7b. The basic element here is comparator C with distinct operating threshold I_t.

A pulse to be measured $I_s(t_0)$; short strobe $I_x(t-\tau)$ of fixed amplitude and duration; and slowly varying feedback signal I_{fb} are simultaneously applied to the comparator. If sum $I_{fb} + I_x + I_s$ exceeds threshold, the comparator switches, generating an output pulse of high fixed amplitude V and appreciable duration T_0. Feedback loop (FB) adjusts current so that the equality

$$I_{fb} + I_x(t_0) + I_s(t_0) = I_t \qquad (2.11)$$

is fulfilled. Then, using signal I_{fb}, one can find I_x at instant $t_0 = t + \tau$ the strobe reaches a maximum. This short strobe is formed by generator G from reference pulse $I_0(t)$ delayed by time. Slowly varying this delay, one can record the waveform of I_x. It should be noted that semiconducting samplers use another, selection/storage, approach to measure weak signals. The operation and performance of semiconductor samplers are detailed in Ryabinin (1968). A superconducting analogue to the selecting/storing device has not been found yet. The aim of this section is to review the advances in superconducting electronics in the field of Josephson comparator samplers. This issue is also topical in the light of discovery of new superconducting compounds, such as HTS (Bednorz and Muller 1986), magnesium diboride MgB_2 (Nagamatsu et al. 2001) and Fe-based superconductors (Askerzade 2012). Contrary

Fig. 2.7 a Block diagram of
comparator sampler **b**
Operation of JJ comparators

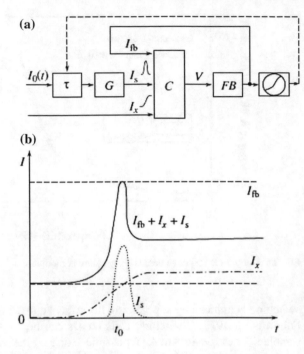

to the early expectations, MgB_2 is now viewed as a superconductor capable of replacing LTS counterparts. To estimate the potential of the new superconductors as the material for samplers, it is necessary to theoretically determine their associated time resolution δt and sensitivity δI_x.

2.2.1 Comparators on Single JJ

Early superconducting samplers were based on two tunnel JJs, one of which served as a generator of short strobes and the other as a comparator. Such a design of the sampler was proposed in Faris (1980). The basic circuit of this comparator is shown in Fig. 2.8. It consists of JJ J_4, which is connected to an interferometer based on two JJs J_1 and J_2 through resistor R and shaper J_3. Switching of the interferometer is accomplished by appropriately selecting feed current I_{feed} and control current I_{cont}. A pulse steep edge that is observed when the interferometer is switched into the resistive state becomes still steeper under the action of small junction J_3. As a result, a short strobe is formed, which is applied to comparator J_4, where it adds up with signal current I_3 and feedback current I_{fb}. Once the critical current has been exceeded, the JJ switches into the resistive state and, accordingly, the feedback current changes.

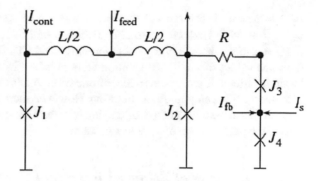

Fig. 2.8 JJ based comparator on J$_4$ (J$_1$ and J$_2$ presents two-junction interferometer, J$_3$ is the sharper for strobe-pulses)

The dynamics of the circuit shown in Fig. 2.8 was simulated and analyzed in detail in paper (Askerzade 2000a). Using the PSCAN (Personal Superconductive Circuit ANalyzer) software package developed in Moscow State University (Polonsky et al. 1991), the shape of the strobe versus feed current I$_{feed}$ and McCumber parameter $\beta_C = \frac{2\pi I_c R_N^2 C}{\Phi_0}$ was studied and strobe rise time isolines on the parameter plane (I$_{feed}$, β_C) were constructed (Askerzade 2000a). It was shown that, for JJs with $\beta_C < 3$, the circuit shown in Fig. 2.8 becomes inappropriate because of a distortion of the strobe shape. The oscillation amplitude at the tail of strobe increases, becoming comparable to the "core" of the pulse (Askerzade 2000a). Such a conclusion comes into conflict with the results of Wolf et al. (1985a), where it was argued that a picosecond time resolution may be reached by further growing of current density, i.e. by lowering β_C.

2.2.2 Time Resolution of a Single-Junction Comparator

The time resolution of the Josephson comparator is fundamentally limited by its inertial properties. According to Likharev et al. (1990), this parameter for a single JJ comparator can be roughly estimated by the expression

$$\Omega_p(0)t = \frac{1}{(1 - \frac{I_t}{I_c})^{1/2}},$$ (2.12)

where

$$\Omega_p^2(0) = \frac{2eI_c}{\hbar C} = \frac{2ej_c t_i}{\hbar \varepsilon \varepsilon_0}.$$ (2.13)

is the plasma frequency of a tunnel JJ (see also Eq. 1.67) (Likharev 1986), t_i is the thickness of an insulating spacer between the superconductors, ε is the permittivity, and ε_0 is the dielectric constant. In expression (2.12), ratio I_t/I_c characterizes the proximity of the real threshold of action of the comparator, to its critical current I_c. Under the actual conditions of thermal fluctuations, $1 - I_t/I_c$ is on the order of

$\gamma_T^{1/3}$, where γ_T is the ratio of thermal energy kT to Josephson energy $E_J = \frac{\hbar I_C}{2e}$, $\gamma_T = \frac{kT}{E_J} = \frac{2ekT}{\hbar I_c}$ (Likharev 1986). For JJs based on LTS, γ_T can be set equal to 0.01. If $V_c = 2\,\text{mV}$, which is typical of commonly used LTS (Nb, Pb), $\delta t = 3\,\text{ps}$. Nearly the same values of the time resolution were obtained in Wolf et al. (1985a) for Pb-based junctions. It was noted in the introduction to this chapter, that a strobe pulses must have a fixed amplitude and duration. However, thermal and fabrication-induced fluctuations in the circuit of a two-junction interferometer make these parameters and, hence, the performance of samplers unstable.

2.2.3 Dynamics of Balanced Josephson Comparator

An important step toward improving the performance of superconducting samplers is to raise their stability against noise and temperature fluctuations. To this end, it was suggested that JJs be configured into a Goto pair to form a balanced scheme (Bakhtin et al. 1983; Kornev and Semenov 1987). Here, two identical junctions are connected in series relative to a strobe signal and in parallel relative to a signal to be measured. The operation of the Goto pair has long been known and is applied, e.g., in semiconductor samplers based on Esaki diodes (Karklinsh and Khermanis 1980). One more advantage of balanced comparators is that distortions due to the nonideality of a strobe pulses are eliminated.

The circuit presented in Fig. 2.8 is easily converted to a balanced circuit if junctions J_3 and J_4 are made identical. To study the dynamics of a balanced comparator on JJs, it is necessary to construct the switching characteristic of the comparator on the plane $(\frac{dI}{d\tau}, I_s)$, where $\frac{dI}{d\tau}$ is the rate of strobe current and I_s is the signal current. The switching characteristic is constructed as follows: as rate of rise $\frac{dI}{d\tau}$ grows, the signal current must be increased so that, for switching time $R_N C$ of the first junction, the current through the second junction does not reach critical value I_c, which means

$$\frac{dI}{d\tau} R_N C < I_s. \tag{2.14}$$

Another way of designing balanced comparators around tunnel JJs is the use of an internal clock generator as a strobe pulse (Askerzade and Kornev 1994b). The dynamics of related processes was numerically simulated with the PSCAN software package (Polonsky et al. 1991). It was found (Askerzade 2000b) that both junctions are switched into the resistive state when the signal is low. The time resolution approaches its fundamental limit if sampling rate is too low, which is observed when $\alpha \beta_C < \delta i$ with δi is the fluctuation smearing of the comparator's threshold characteristic. It is well known that fluctuation smearing $\delta i = \frac{I_T}{I_c}$ (Likharev 1986), where critical current I_c varies usually between $10\,\mu\text{A}$–$1\,\text{mA}$ for tunnel junctions in LTS. At liquid-helium temperature, $I_T \approx 0.3\,\mu\text{A}$; for liquid-nitrogen temperature, I_T may reach $32\,\mu\text{A}$.

The dynamics of a two-junction interferometer can be used to reduce distortions in the shape of the strobe pulse and obtain very short single flux quantum (SFQ) pulses. Taking I_{feed} (Fig. 2.8) below the threshold value and appropriate I_{cont}, one can generate a SFQ pulse of area

$$\int V dt = \Phi_0 = 2\,\text{mV} \times \text{ps} \qquad (2.15)$$

amplitude of about $2V_c$, and duration of $\frac{\pi}{\omega_c}$. Such an approach, however, requires that a latching source be used. In this case, it is convenient to apply a generator of RSFQ pulses that is built around a single JJ with a hysteresis-free IV curve (Kornev and Semenov 1987; Gudkov et al. 1988).

2.2.4 Time Resolution of Balanced Tunnel Josephson Comparators

As for any sampler, the time resolution of the given device can be determined using the transfer characteristic $H(\tau)$ (Karklinsh and Khermanis 1980), which relates the output signal of a sampler, V_{out}, to a signal applied to the input of the comparator in the form of a small step, $I_s = I\theta(t)$. The time resolution is defined as the time taken for the characteristic $H(\tau)$ to rise from 10 to 90% of its maximal value, where τ is the delay (advance) time of the strobe pulses relative to the time the step is applied to the input of the comparator. For identical JJs J_3 and J_4 (Fig. 2.8), the following phase relationships are valid:

$$\beta_C \ddot{\phi}_3 + \dot{\phi}_3 + \sin\phi_3 = i_0, \qquad (2.16)$$
$$\beta_C \ddot{\phi}_4 + \dot{\phi}_4 + \sin\phi_4 = i_0 + i_1. \qquad (2.17)$$

Here, i_0 and i_1 are the currents in units of the JJ critical current (i_0 is the strobe-induced current, and i_1 is the sum of the feedback and signal currents); τ is given in units of $\Phi_0/2\pi I_c R_N$; Φ_0 is the flux quantum, and R_N is the JJ resistance in the normal state. For small deviations of the phase relative to the strobe pulses, $\delta\phi = \phi_3 - \phi_4$, we have

$$\beta_C \ddot{\delta\phi} + \dot{\delta\phi} + \cos\left(\frac{\phi_+}{2}\right)\dot{\delta\phi} = i_1, \qquad (2.18)$$

$$\beta_C \frac{\phi_+}{2} + \frac{\phi_+}{2} + \sin\frac{\phi_+}{2} = i_0, \qquad (2.19)$$

where $\phi_+ = \phi_3 + \phi_4$. An asymptotic solution of Eq. (2.19) in the case of a linearly growing sampling current through a JJ, $i_0 = \alpha\tau$, is given in Likharev (1986) and Snigirev (1984) (In this section α denotes the sampling rate.)

$$\frac{\phi_+}{2} = \frac{\pi}{2} + \left\{ \begin{array}{l} -\sqrt{-2\alpha(\tau - \alpha^{-1})}, \text{ at } \phi_+ \to -\infty \\ u_0(\tau - \alpha^{-1} - \tau_0) \text{ at } \phi_+ \simeq 0 \\ \frac{12\beta_C}{(\tau-\alpha^{-1}-\tau_d)^2} \text{ at } \phi_+ \to \infty \end{array} \right\}, \tag{2.20}$$

where $u_0 = 1.64\alpha^{3/5}\beta_C^{-1/5}$ and $\tau_0 = 0.95\alpha^{1/5}\beta_C^{-1/2}$. The mean delay time τ_d is given by

$$\tau_d = 4.64\alpha^{-1/5}\beta_C^{2/5}. \tag{2.21}$$

Let $i_s = \Theta(t - \tau)$ be the input signal in the form of Heaviside step function at time instant $t = \tau$. Then, the solution to Eq. (2.18) has the form

$$\delta\phi = \int_{-\infty}^{t} K(t, \xi)\left[\Theta(t - \xi) + i_{fb}(\tau)\right]d\xi. \tag{2.22}$$

If $\beta_C\alpha \ll 1$, the kernel in case of a linearly growing current, $i_0 = \alpha\tau$, has the form

$$K(t, \xi) = \pi Ai(G\xi)Bi(Gt), \tag{2.23}$$

where $Ai(\ldots)$ and $Bi(\ldots)$ are the Airy functions and $G = 1.14^{1/3}\alpha^{1/5}\beta_C^{-1/5}$. For any τ, transfer characteristic $H(\tau)$ obtained from the condition of nongrowing solution has the form

$$H(t) = H_0^{-1}\int_{-\infty}^{t} d\xi Ai(G\xi); \quad H_0 = \int_{-\infty}^{\infty} dt Ai(Gt). \tag{2.24}$$

Figure 2.9 shows that the transfer characteristic of a balanced comparator based on tunnel JJs is of oscillatory character at $\tau < 0$. Such behavior is associated with a complete set of reactances in the equivalent circuit of a JJ at $\beta_C \gg 1$ (see Fig. 1.19) (Likharev 1986). From Fig. 2.9, we find that the time resolution of tunnel JJ

Fig. 2.9 Transfer characteristic of a balanced tunnel JJ comparator

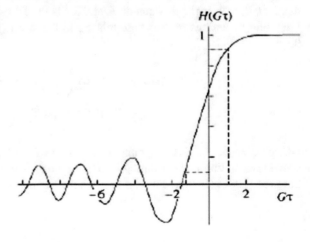

comparators is

$$\delta t = 2.3\alpha^{-1/3}\beta_C^{2/5}. \tag{2.25}$$

As follows from last equation, time resolution δt depends on capacitance β_C of a JJ and sampling rate α of the strobe-induced current through the junction. A pair of tunnel JJs ceases to operate as a comparator when $\alpha\beta_C < \frac{I_T}{I_c}$. Under the assumptions made above, the time resolution is estimated at a level of several tens of picoseconds. This means that, for slow strobes, the time resolution is far from the limiting value of the superconducting devices (Wolf et al. 1985a). Thus, it is necessary either to reject the balanced tunnel scheme of a Josephson comparator and use unbalanced designs or to design balanced comparators on hysteresis-free (overdamped) JJs, which are considered next section.

2.2.5 Balanced Comparators on Overdamped JJs

Figure 2.10 shows the equivalent circuit of a development a device that handles RSFQ idea (Kornev and Semenov 1987; Gudkov et al. 1988). JJ J_1 shunted by resistor R_1 generates flux quantums with a repetition rate that is proportional to mean voltage V_1 across this junction and is controlled by bias current I_1, which far exceeds its critical current I_c. RSFQ pulses acting as strobes enter a buffer discrete transmission line made of JJs J_2–J_4. The output of this line is connected to a balanced comparator on Josephson pair J_6 and J_7 through junction J_5. Bias currents I_2 and I_3 of the Josephson transmission line (JTL) and comparator are lower than the critical current of these junctions. When such a device is used as the pulsed comparator of a sampler, the output signal is mean voltage V_{av} across JJ J_7 (or J_6), which is proportional to the probability that a single flux quantum escapes loop J_4–J_7 precisely from this junction. Current I_4 is a sum of the current to be measured, I_s, and feedback current I_{fb}. The feedback makes current I_4 equal to threshold current I_t when the probability of a RSFQ pulses escaping is 0.5. The output voltage of the sampler, V_{out}, is proportional

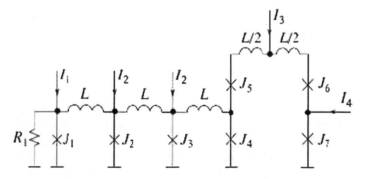

Fig. 2.10 Equivalent representation of RSFQ circuit developed in Gudkov et al. (1988)

to I_{fb}. The dynamics of a balanced comparator on overdamped JJs was analyzed in detail in papers (Gudkov et al. 1988; Snigirev 1984; Askerzade et al. 1990; Askerzade and Kornev 1994a).

2.2.6 Time Resolution of Balanced Overdamped Josephson Comparators

Here, we will use the same dynamic equations (Eqs. (2.16) and (2.17)) as for tunnel junctions. Since the McCumber parameter is small, $\beta_C \ll 1$, the first terms in these equations are ignored. For the circuit in Fig. 2.10, the indice for the Josephson phases are changed and, for a linearly growing current through a overdamped junction, we have

$$\frac{\phi_+}{2} = \frac{\pi}{2} + \left\{ \begin{array}{ll} -\sqrt{-2\alpha(\tau - \alpha^{-1})}, & \text{at } \phi_+ \to -\infty \\ u_0(\tau - \alpha^{-1} - \tau_0) & \text{at } \phi_+ \simeq 0 \\ \frac{2^{1/3}}{(\tau - \alpha^{-1} - \tau_d)^{1/3}} & \text{at } \phi_+ \to \infty \end{array} \right\}, \tag{2.26}$$

here $u_0 = 1.64\alpha^{3/5}$ and $\tau_0 = 1.21\alpha^{-1/3}$. The mean delay time τ_d is given by

$$\tau_d = 2.9\alpha^{-1/3}. \tag{2.27}$$

Using the solution

$$\delta\phi = \int_{-\infty}^{t} \exp\left(-\frac{\phi_+(x)dx}{2}\right) \int_{-\infty}^{t} \left[\Theta(t - \xi) + i_{fb}(\tau)\right] d\xi. \tag{2.28}$$

and the Eq. (2.26), we obtain, upon integration, an expression for the transfer characteristic $H(\tau)$

$$H(\tau) = 1 - \text{erf}\left\{\sqrt{\frac{1.25}{2}}\alpha^{1/3}\tau\right\} \tag{2.29}$$

where erf(...) is the error integral. Then, using tabulated data for the error integral, one can estimate the time resolution of balanced overdamped comparators as

$$\delta t = 0.64\frac{\Phi_0}{2\pi I_c R_N}. \tag{2.30}$$

When deriving of expression (2.30), we assumed that a SFQ pulse has a triangular shape and so the sampling rate can be written as $\alpha = 2/\pi$. For real values of the characteristic voltage $V_c = I_c R_N = 2\,\text{mV}$, the time resolution is about $1\,\text{ps}$. The above analysis is referred to the case when a pulsed comparator is based on a Goto pair. Actually, however, the behavior of this system is more complicated (Askerzade and Kornev 1994a); specifically, the effect of inductance in the circuit is disregarded in the analysis given above. Therefore, the process dynamics in this circuit was numerically simulated in the absence of fluctuations (Askerzade and Kornev 1994a) in order to see how phases ϕ_6 and ϕ_7 of respective JJs of the balanced comparator vary when a RSFQ pulses transferring the comparator to an unstable state at $I_4 = I_t$ (in the absence of signal I_s to be measured) comes to loop J_4–J_7. The results of this simulation at different values of I_3 for the case when either both junctions ($I_t = 0$) or one ($I_t \neq 0$) junction is in an unstable state are shown in Fig. 2.11a, b, respectively. Now suppose that a single current step, $I_s = 1(t - \tau)$, is applied to the input of the comparator at time instant $t = \tau$. For a small deviation of the phase from an equilibrium value in both unstable state of the comparator, one can write a linear equation with time-varying coefficient $G(t)$ (Askerzade and Kornev 1994a)

$$\dot{\phi}_3 + G(t)\phi_3 = \Theta(t - \tau) + i_{fb}, \qquad (2.31)$$

where $G(t) = \cos \phi_6(t)$. In the first case of unstable equilibrium of the comparator ($I_t = 0$), phase ϕ is the difference between the phases of the JJs,

$$\phi = \phi_-(t) = \phi_6(t) - \phi_7(t). \qquad (2.32)$$

Fig. 2.11 Dynamics of the circuit implemented in Gudkov et al. (1988) $I_3/I_c = 0.8$ (a) and 1.4 (b). Other parameters: $l = 0.4$; $I_1/I_c = 0.7$; $\beta_C = 0.1$; time in units $\hbar/2eV_c$

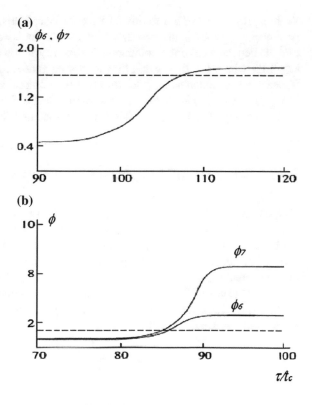

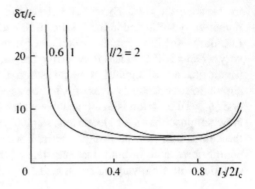

In the second case, when $I_t \neq 0$, phase ϕ is the difference between the actual value of phase ϕ_6 from the value of ϕ_6 calculated above (Fig. 2.11b). For the transfer characteristic, we have

$$\delta\phi = -i_{fb} = H_0^{-1} \int_\tau^\infty \exp\left(-\int_\xi^\infty G(x)dx\right) d\xi; \quad H_0 = \int_{-\infty}^\infty \exp\left(-\int_\xi^\infty G(x)dx\right) d\xi.$$
(2.33)

With expression (2.33), a family of transfer characteristics for different values of parameters $i_3 = I_3/I_c$ and $l = 2\pi LI_c/\Phi_0$ was calculated. This made it possible to find the dependence of the comparator's time resolution on these parameters. From Fig. 2.12, where the time resolution is plotted versus $I_3/2I_c$ for various values of dimensionless inductance l, it follows that the minimal value of the time resolution is $\delta\tau_{min} = 5t_c = 5\hbar/2eV_c$. For characteristic voltage $V_c \approx 2\,\mathrm{mV}$ commonly used in the technology of LTS, $\delta\tau_{min} = 0.8\,\mathrm{ps}$ (Askerzade and Kornev 1994a).

2.2.7 Sensitivity of Josephson Comparators

The sensitivity of any comparator, can be determined from the switching characteristic. In the absence of fluctuations, this characteristic has a threshold character: the mean voltage across the junctions as a function of the signal amplitude has the form of an ideal step. The sensitivity of a comparator depends on a fluctuation-induced smearing of this characteristic. When the strobe pulse is rather wide, i.e., when $\alpha \ll (3\gamma_T/2)^{2/3} < 1$ (α is the dimensionless sampling rate), the smearing of the threshold characteristic of any (tunnel or hysteresis-free) JJs depends on thermal fluctuations (Askerzade 1998)

$$\frac{\Delta I_x}{I_c} \equiv \gamma_T^{2/3}$$
(2.34)

As follows from Eq. (2.34), the final result is independent of McCumber parameter β_C. This is because the Josephson phase dynamics is the same at all junctions when the sampling rate is low (Likharev 1986; Snigirev 1984). However, expression (2.34) fails in the limit of SFQ pulses, since the rate of growing of the current, α, increases in this approximation, exceeding $(3\gamma_T/2)^{2/3}$. The dynamics and fluctuation of the delay time in this limit were studied in papers (Askerzade 1998, b). Here, of most importance is the fact that the sensitivity starts depending on rate of growing α and McCumber parameter β_C

$$\frac{\Delta I_x}{I_c} \equiv \begin{cases} 0.17\sqrt{\gamma_T}(\frac{\alpha}{\beta_C})^{3/8}, & \beta_C \gg 1 \\ 3.44\sqrt{\gamma_T}\alpha^{7/18}, & \beta_C \ll 1 \end{cases}. \tag{2.35}$$

From Eq. (2.35), it follows that, when α is high, the dependence of ΔI_x on thermal fluctuation intensity γ_T obeys a square-root law whatever junction capacitance β_C. The fact is that, at a high rate of growing, fluctuations somewhat change the switching process. In both limits ($\beta_C \gg 1$ and $\beta_C \ll 1$), the threshold switching characteristic smears in proportion to rate of rise α unlike time resolution δt. On the other hand, the smearing of this characteristic in the case of tunnel junctions is insignificant compared with overdamped junctions, since the inertia of the former is high. Because of high $\beta_C \gg 1$, the basic stage of switching takes an extremely short time; i.e., having gained a high velocity, a "particle" corresponding to the Josephson phase, moves "by inertia". In general, sensitivity δI_x of comparators is directly proportional to smearing ΔI_x of the threshold switching characteristic

$$\delta I_x = \frac{\Delta I_x}{\sqrt{N}}. \tag{2.36}$$

In this expression, N is an averaging factor related to the statistical character of measurement. It is equal to the number of pulses used to measure an output pattern at a time point. The value of N considerably depends on pulse repetition rate f. The maximal duration of a signal to be measured is expressed as $\tau_{max} \leq f^{-1}$. This means that the formula

$$N = \frac{T}{\delta t}\tau_{max}f \tag{2.37}$$

is valid if the total time of measurement of the pulse is T.

From the expressions for the time resolution, it follows that δt depends on characteristic voltage V_c. According to the microscopic theory (see for example Likharev 1986; Barone and Paterno 1982), the characteristic voltage is determined by the order parameter of the superconducting electrodes at a given temperature. For the highest performance LTS, $V_c \approx 2\,\mathrm{mV}$, which corresponds to time resolution $\delta t = 0.1\,\mathrm{ps}$ in the case of RSFQ pulses. For well-known Nb based JJs, δt is about 1.6 ps (600 GHz) (Gubankov et al. 1983). In junctions based on HTS with critical temperature $T_c = 90\,\mathrm{K}$, and order parameter $\Delta_0/e = 20\,\mathrm{mV}$, characteristic voltage V_c is expected to be several millivolts at liquid nitrogen temperature $T = 77\,\mathrm{K}$.

Bicrystalline junctions on $SrTiO_3$ substrates were reported in Vale et al. (1997). The possibility of increasing characteristic voltage V_c to 1 mV was discussed in Poppe et al. (2001) and Divin et al. (1997). Recently in Borisenko et al. (2005), JJs have been prepared on YBCO superconductors. Here, critical current density j_c reach $(2 - 5) \times 10^5 \, A/cm^2$ at liquid nitrogen temperature and characteristic voltage is $V_c = 0.6 - 0.9 \, mV$. The hysteresis-free IV curve of JJs on HTS makes these materials promising for designing RSFQ systems (see Chap. 3).

A HTS balanced comparator on JJs was first implemented in Sonnenberg et al. (1999). The operating frequency of that device was 72 GHz, which corresponds to a time resolution of 14 ps. In Saitoh et al. (2002), an YBCO balanced comparator was used as a component of a delta–sigma modulator. The parameters of this device were measured at an operating temperature of 20 K and an operating frequency of 100 GHz. The first YBCO sampler was implemented and then improved by Japanese scientists (Hidaka et al. 2001; Maruyama et al. 2003). The operating frequency of this sampler was equal to 120 GHz. According Eq. (2.36), sensitivity δI_x of Josephson comparators depends on the smearing of the threshold switching characteristic, ΔI_x, and parameter N. The value of N strongly depends on pulse repetition rate f. In modern superconducting electronics (Likharev 2002), f on the order of 100 GHz and $\tau_{max} \approx 10 \, ps$ may be attained using electrically controlled delay line based on JJs. In this case, $f\tau_{max} \approx 1$. Putting $T_m = 10^{-3} \, s$ and $\delta t \approx 1 \, ps$, we obtain $N \approx 10^9$. The "radiometric gain" in expression (2.37) is then $\sqrt{N} \simeq 3 \times 10^4$. For LTS, ΔI_x varies from 10 μA to 1 mA, which corresponds to sensitivity δI_x varying from 0.3 nA to 0.3 μA. For HTS at liquid nitrogen temperature, we have $\delta I_x = 1 \, nA$.

As the pulse repetition period grows, the radiometric gain decreases as (here, it is more convenient to consider τ_{max}). Yet, the sensitivity remains high. The value $\delta I_x = 1 \, nA$ is better than that of semiconductor samplers. The smearing of the threshold characteristic of a balanced comparator on niobium junctions was measured and analytically calculated in recent study (Walls et al. 2002) on quantum fluctuations in these devices. The data calculated in terms of the Calderia–Leggett theory (Caldeira and Leggett 1983a) were compared with the temperature dependence of smearing ΔI_x in the interval 1.5–4.2 K. It was found that, at higher temperatures, ΔI_x grows as $\sqrt{T}$, in accordance with expression (2.35). However, as the temperature tends toward zero, ΔI_x tends to saturate because of quantum fluctuations. It is also remarkable that ΔI_x depends on parameter β_C in the quantum limit. Contrary to the prediction by expression (2.35), smearing ΔI_x increases with β_C at $T \to 0$. Thus, balanced Josephson comparators offer a powerful means for studying quantum fluctuations in such systems (Caldeira and Leggett 1983a) and seem to be promising for investigating the properties of qubits. In closing, efforts to create LTS samplers using latching technology date back to the 1980s (Wolf et al. 1985a; Faris 1980). Hypres Co. (United States) even launched Josephson stroboscopic oscilloscopes–reflectometers in serial production (Whiteley et al. 1987). However, samplers on LTS have not found wide application, possibly because of the need to cool down the system to helium temperatures. At the same time, Japanese researchers guess that samplers made of HTS may have extensive applications, because the problem of cooling down to cryogenic temperatures here is lacking (unlike in the technology of LTS). Application of fairly cheap many-pass nitrogen coolers to a great extent eliminates cooling-related difficulties. Note that creating superconducting samplers is also dictated by recent advances in computing facilities. These devices seem to fit naturally into digital (both semiconducting and superconducting) circuits operating at nitrogen temperatures.

2.3 Superconducting Interfometers and SQUIDs

This section is concerned with the basic principles underlying the devices knows as SQUIDs. It is well known that such devices have opened new horizons in LTS measurement technique. A SQUID is the most sensitive measurement device which can measure magnetic flux on the order of one flux quantum Φ_0. The magnetic properties of different systems including spin, superspin glass, and superparamagnet are studied using SQUID devices. Firstly we will discuss flux quantization in a superconducting ring with one and two JJ (single and two-junction interferometers). These interferometers used in SQUIDs as sensitive elements. Below we present details of physical foundation for operation and sensitivity of SQUIDs.

2.3.1 Flux Quantization in a Superconducting Ring with Single JJ

2.3.1.1 Basic Equations and the Case of 0 and π Junctions

The basic element of a single junction interferometer is a superconducting ring containing JJ. In general, the flux on the JJ Φ is not equal external magnetic flux Φ_e. Their difference is due to the screening current circulating in the superconducting ring (Fig. 2.13):

$$\Phi = \Phi_e - LI_c \sin \phi. \tag{2.38}$$

For dimensionless notation we have

Fig. 2.13 Single junction interferometer

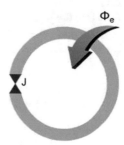

Fig. 2.14 $\Phi(\Phi_e)$ dependence of single junction interferometer for different dimensionless inductance $l = \frac{2\pi L I_c}{\Phi_0}$

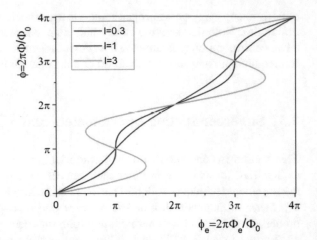

$$\phi = \phi_e - \ell \sin \phi, \tag{2.39}$$

where external flux ϕ_e in units Φ_0. $\phi_e = \frac{2\pi \Phi_e}{\Phi_0}$, dimensionless inductance $\ell = \frac{2\pi L I_c}{\Phi_0}$. Dependence of $\phi(\phi_e)$ for different dimensionless inductivity parameters ℓ presented in Fig. 2.14. Detail analysis behavior of single junction interferometer with junction based on s-wave superconductors and their dynamics presented in Likharev (1986), Barone and Paterno (1982).

The most interesting effect associated with π junction is that its sign changes in the critical current I_c of JJ, suggested by Sigrist and Rice (1992), Bulaevskii et al. (1977). It was shown that a superconducting ring containing an odd number of π shifts will (under certain conditions) spontaneously generate a magnetic flux $\Phi_0/2$. Flux quantization of a multiply connected superconductor represents one of the most fundamental demonstrations of macroscopic phase coherence in the superconducting state. For a ring with an odd number of sign changes in the circulating supercurrent I_s, it is sufficient to consider the case in which only one critical current is negative (say, $I_s = I_c \sin(\phi + \pi)$). In the absence of applied magnetic field, the ground state of a superconducting ring containing an odd number of sign changes (π ring) has a spontaneous magnetization of a half magnetic-flux quantum (i.e., $I_s L = \frac{\Phi_0}{2}$). If the ring contains an even number of π shifts, including no π shifts at all (zero ring), $I_s = 0$ in the ground state, and the magnetic-flux state has the standard integral flux quantization ($\Phi = n\Phi_0$). Free energy of superconducting ring in general case has a form

$$U(\Phi, \Phi_e) = U_L + U_S = \frac{\Phi_0^2}{2L} \left\{ (\phi - \phi_e)^2 - \frac{L I_c}{\pi \Phi_0} \cos(\phi + \varphi) \right\}, \tag{2.40}$$

where $\varphi = 0, \pi$ for the zero ring and a π ring, respectively. It is well known that conventional superconductors generally tend to expel a small external magnetic field under transition into the superconducting state. This "Meissner effect" leads to complete or (due to remnant trapped flux, e.g., in polycrystalline superconductors) partial diamagnetism. Paramagnetic "Meissner effect" was observed in ceramic $Bi_2Sr_2CaCu_2O_8$, which can be explained by the d-wave character of superconducting order parameter. Thus, d-wave symmetry of order parameter leads to frustrated Josepshon junction circuits (π rings) in polycrystalline superconductors, where orientation of grains in sample seems random and contact each other.

Spontaneous generation of a with a flux $\Phi_0/2$ at the meeting point of Josephson coupled superconducting crystals with unconventional character of pairing in a frustrated geometry was first proposed by Geshkenbein and co-workers (1986). The first experimental observation of this effect in a controlled geometry was in study (Tsuei and Kirtley 2000). In the tricrystal experiment of Tsuei and Kirtley (2000), an epitaxial YBCO film (1200 A^o thick) was deposited using laser ablation on a tricrystal (100) $SrTiO_3$ substrate with the configuration shown in Fig. 2.15. A high-resolution scanning SQUID microscope was used to direct measurement of the magnetic flux threading through HTS rings. The measurement SQUID's used were based on LTS $Nb-AlO_x-Nb$ trilayer SQUID's. Fig. 2.16 shows a scanning SQUID microscope image (Tsuei and Kirtley 2000) of a three-junction YBCO ring in the original tricrystal magnetometry experiments. Sample was cooled to 4.2 K and imaged in a magnetic field estimated to be less than 0.4 μT. It was shown (Fig. 2.17) that the $\Phi_0/2$ flux effect at the tricrystal meeting point persists from T=0.5 K through

Fig. 2.15 Experimental geometry for study of half-flux quantum state of single junction interferometer

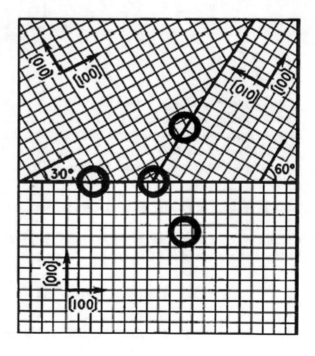

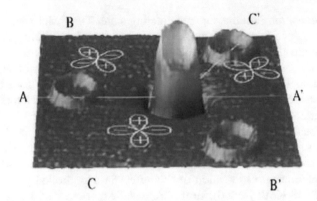

Fig. 2.16 Three-dimensional visualization of SQUID microscope image of a thin-film YBCO tricrystal ring sample, cooled and imaged in nominally zero magnetic field (Kirtley 2010)

Fig. 2.17 Frustrated state of tricrystal geometry as function of temperature (Kirtley 2010)

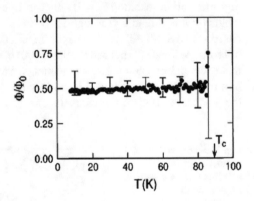

T_c =90 K) with no change in total flux ($\Phi_0/2$). It means that d-wave symmetry of superconducting state dominates in HTS (Kirtley 2010). This implies that the three outer control rings have no magnetic flux trapped in them, but in the three-junction ring at the center has $\Phi_0/2$ total flux. Visualization of the control rings are related with a slight change in the inductance of the SQUID.

2.3.1.2 Effects of Fluctuations on a Single-Junction Interferometer

Calculation of average value of the Josephson term $\langle U_S \rangle$ in Eq. (2.40) for $\phi_e = 0$ yielded to Khlus and Kulik (1975)

$$\langle U_S \rangle = -E_J \cos \langle \phi \rangle \exp \left(-\frac{L}{L_{F0}} \right), \tag{2.41}$$

where L_{F0} is the so called "thermal fluctuational" inductance determined as

$$L_{F0} = \left(\frac{\Phi_0}{2\pi}\right)^2 \frac{1}{kT}. \tag{2.42}$$

For $L > L_{F0}$, the Josephson interference is sharply suppressed by thermal fluctuations. This fact imposes limitations on the ring size in a superconducting single junction interferometer. For helium temperatures it corresponds to the value about 2 nH, which restrict using superconducting rings about 1 mm. At very low temperatures, an important role belongs to quantum fluctuations. Influence of quantum fluctuations on single-junction interferometer considered in Askerzade (2005a). For low inductance interferometers ($\ell \ll 1$), the matrix element of the superconducting current in the ground state is given by the expression

$$\langle 0|I_c \sin \phi|0\rangle = -\frac{2\pi E_J}{\Phi_0} \sin \phi_e \, exp(-L/L_{FQ}), \tag{2.43}$$

where $L_{FQ} = (\frac{\Phi_0}{\pi})^2 \frac{1}{\hbar\omega}$ and the frequency ω is determined as $\omega = \frac{1}{\sqrt{LC}}$, where L is the inductance of superconducting loop, C is the capacitance of JJ. It means that above presented result given by expression (2.41) remains in power in the case of low inductance interferometers, however by replacing the thermal energy kT with the energy of quantum $\hbar\omega$. In the case of a high inductance at low temperatures as in a single junction Josephson interferences is suppressed (Askerzade 2005a).

2.3.2 Basic Schema of AC SQUID and Their Parameters

Schematic representation of AC SQUID presented in Fig. 2.18. A current of frequency ω is supplied by a high-impedance AC current source to the resonance circuit (Zimmerman et al. 1970) and (Mercereau 1970). Detailed review of of operation of

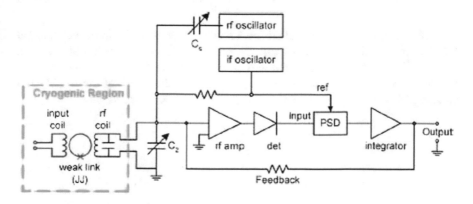

Fig. 2.18 Basic schema of AC SQUID

AC SQUIDs have been written by many authors (Danilov et al. 1980; Clarke 1996; Chesca 2004). Frequencies ω is usually 10–30 MHz, however there are AC SQUIDs operating at GHz. The coil of the resonant circuit is coupled to the superconducting loop of a single junction interferometer. The AC voltage across the tank circuit is magnified by the amplifier. The device input is the coil inductively coupled to the SQUID loop. Historically, it appears that most LTS AC SQUIDs were operated in the hysteretic mode, although it was shown that, there are advantages to the nonhysteretic mode. For this reason, the importance of the nonhysteretic AC SQUID was not widely exploited experimentally. In relation with discovery of HTS has changed this situation dramatically, due to the systematic experimental study of the Germany group from Julich.

2.3.2.1 Hysteretic Mode

First important characteristic of the AC SQUID is the transfer coefficient H of the applied flux Φ_x on the output voltage V. As shown by theoretical analysis in study (Danilov et al. 1980), in the case of hysteretic mode, i.e. high inductance ($\ell \gg 1$) inferferometers transfer coefficient H for AC SQUIDs can be written as

$$H = \left| \frac{\partial V}{\partial \Phi_x} \right| = \frac{\omega_T}{k} \sqrt{\frac{L_T}{L}}; \ k = \frac{M}{\sqrt{LL_T}}, \tag{2.44}$$

where L_T is the inductance of tank circuit and ω_T is a tank frequency. In last expression inductive coefficient k should be greater than $Q_T^{-1/2}$, where Q_T is the quality factor of the tank circuit (see Fig. 2.18). Estimation of transfer coefficient H for AC SQUIDs leads to results 10 μV/Φ_0. Another characteristic of the SQUIDs is the energy sensitivity, which is determined as

$$E_V = \frac{(\delta \Phi_x)^2}{2L \Delta f}, \tag{2.45}$$

where $\delta \Phi_x$ is the magnetic flux at the level noise of SQUID in the frequency interval Δf. Estimation of energy sensitivity E_V of AC SQUID gives result at the level 5×10^{-29} J/Hz, in which we take into account for interferometer inductance the value $L = 3 \times 10^{-10}$ H. The corresponding intrinsic flux noise of the AC SQUID calculated in the study (Kurkijarvi 1973) and is given by the expression

$$S_\Phi^i(f) = \frac{(LI_c)^2}{\omega_T} \left(\frac{2\pi kT}{I_c \Phi_0} \right)^{4/3}. \tag{2.46}$$

2.3.2.2 Nonhysteretic Mode

For low inductance interferometers ($\ell \ll 1$) the operation of the nonhysteretic AC SQUID were described (Likharev 1986; Hansma 1973; Ryhanen et al. 1989). For low inductance interferometers ($\ell \ll 1$), the total magnetic flux threading the SQUID is nearly equal to the applied flux, and $\omega_T = \frac{I_c R}{\Phi_0}$. The transfer coefficient in this regime can be written as

$$H = \frac{2}{\pi}\kappa^2 Ql \frac{\omega_T}{k} \sqrt{\frac{L_T}{L}}. \tag{2.47}$$

This transfer coefficient exceeds that of the hysteretic AC SQUID by a factor of order $\kappa^2 Ql$, which can be made larger than unity for the nonhysteretic case by choosing $\kappa^2 Q \gg 1$. The intrinsic noise energy of LTS, nonhysteretic, AC SQUIDs has been calculated by several authors, and is approximately

$$E_V = \frac{3kT}{\ell^2 \omega_c}. \tag{2.48}$$

In derivation of last expression the external frequency is set equal to cutoff frequency of SQUID $\omega_c = L/R_N$ and thermal fluctuations considered as perturbation $\gamma_T = \frac{kT}{E_J} \ll 1$. This is generally much lower than for the hysteretic mode. An energy sensitivity in this mode at the level of $20\hbar$ (Kuzmin et al. 1985). The intrinsic energy sensitivity remains low even in the large fluctuation $\gamma_T > 1$. Optimized value of dimensionless inductance is $\ell = 1/\gamma_T$ (Chesca 1998). As a result, nonhysteretic AC SQUID can be operated with relatively large inductance and hence large effective area, increasing their sensitivity.

Value of energy sensitivity E_V for different types of SQUIDs is considerably different. AC SQUIDs are characterized by the high value of noise sensitivity E_V. AC SQUIDs has simple construction in comparison with DC SQUID. However, using AC current for measurements causes additional complication in application. High value of E_V is not determined by fundamental fluctuations in single junction interferometer and depends noises of AC SQUID amplifier (Koelle et al. 1999). The overall noise energy of the hysteretic AC SQUID should not increase very much as the temperature is raised from 4 K to 77 K. However, it should be noted that there are no simulations or calculations to the best of our knowledge in this respect. In the case of DC SQUID, the overall noise energy will increase significantly as the temperature is raised to 77 K (Koelle et al. 1999) since a properly designed circuitry is limited largely by intrinsic noise at 4.2 K. The energy sensitivity of different types of SQUIDs as function of frequency presented in Fig. 2.19 (Fagaly 2006).

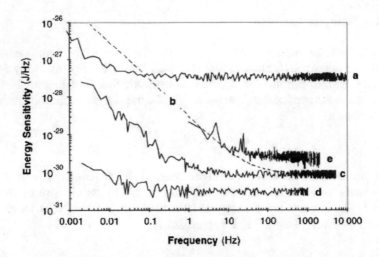

Fig. 2.19 Energy sensitivity of SQUIDs versus frequency for different SQUID device. a is a LTS AC SQUID operated at a bias frequency of 19 MHz; b is a DC biased LTS DC SQUID with amorphous silicon barriers; c is b using AC biasing; d is a DC biased LTS DC SQUID with AlOx barriers; and e is an AC biased HTS DC SQUID utilizing a ramp edge junction. Devices a–d were operated at 4.2 K; device e was at 77 K (Fagaly 2006)

2.3.3   Double Junction Interfometer and DC SQUID

2.3.3.1   JJs Based on S-Wave Superconductors

Two-junction interferometer consists of two JJs in parallel, forming a superconducting ring (Fig. 2.20). Suppose that a magnetic flux Φ passes through the interior of the loop. Equations describing the double junctions interferometer dynamics can be regarded as an equation of motion of a point mass in a field of force with a 2D potential

Fig. 2.20 Double junction
interferometer

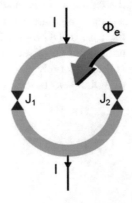

$$\frac{U(\phi_1, \phi_2)}{2E_J} = \frac{\pi}{\ell} \left(\frac{\phi_1 - \phi_2}{2} - \frac{\Phi}{\Phi_0} \right)^2 + \frac{i}{2}\pi \frac{\phi_1 + \phi_2}{2} - \cos \pi \frac{\phi_1 + \phi_2}{2} \cos \pi \frac{\phi_1 - \phi_2}{2},$$

$$(2.49)$$

where ℓ is the normalized total inductance of double-junction interferometer, ϕ_i $i = 1, 2$ is the phase of JJ (J_1 and J_2). As shown by simulations, for $i = 0.4$, $U(\phi_1, \phi_2)$ has multiple metastable state separated by saddle points on the $\phi_2 = \phi_1$ line. By increasing bias current I, these saddle points gradually disappear. At $I > I_c$ it seems that all the saddle points disappear, suggesting no stable state corresponding to local minima of the potential energy. Detail description of double-junction interferometers on conventional superconductors were presented in Likharev (1986) and Barone and Paterno (1982).

In the case of double-junction low inductance interferometers ($\ell \ll 1$) total current can be written as

$$I = 2I_c \sin(\phi_1 - \pi\phi_e) \cos \pi\phi_e. \tag{2.50}$$

It means, that the critical current of double junction interferometer in this limit is a periodic function of the external magnetic flux. Ideal case for the I/I_c versus Φ_{ext}/Φ_0 curve in the double junction interferometer, where I is the maximum supercurrent $I_{max} = 2I_c |\cos \pi\phi_e|$. $I = 2I_c$ when $\Phi_{ext}/\Phi_0 = n$ (integer) and $I = 0$ for $\Phi_{ext}/\Phi_0 = n + 1/2$. Voltage variation across the two-junction interferometer resulting from modulation of the maximum zero-voltage current by an externally applied flux through the superconducting loop also presented in Fig. 2.21.

Fig. 2.21 IV curve of two junction interferometer under external magnetic field

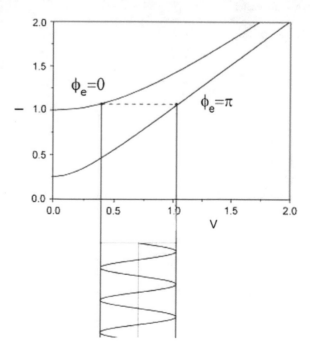

2.3.3.2   JJs Based on D-Wave Superconductors

The predominantly $d_{x^2-y^2}$ symmetry of the order parameter (see Chap. 1) in HTS influences fundamental properties of JJs and superconducting interferometers. We consider two-junction π interferometers which consist of one JJ with negative critical current (Harlingen 1995). In papers (Chesca et al. 2002; Chesca 1999), the theories describing the effects of LC resonances on the IV curve of conventional interferometers have been extended, accounting for a possible complex mixed symmetry $\varepsilon s + i(1 - \varepsilon)d_{x^2-y^2}$, $0 < \varepsilon < 1$ of the order parameter. Experimental realization and notation of a π two-junction interferometer presented in Fig. 2.22 (Chesca et al. 2002). The detail analysis (Chesca 1999) predict $d_{x^2-y^2}$-wave induced zero-field current enhancements due to resonances to appear in the IV curve of π interferometers with small JJs (Chesca et al. 2002). The self-induced resonances occurring in two-junction π interferometers have been observed at temperatures of 4.2–77 K (Fig. 2.23) (Chesca et al. 2002). These resonances reveal that the $d_{x^2-y^2}$ wave symmetry induces in zero applied magnetic field circulating AC currents, which oscillate with Josephson frequency ω_J. The high frequency behavior of π interferometers provides an evidence to complement the behavior of ordinary two-junction interferometer. Experimentally proved that the π shift observed before in the zero-voltage state is present for Josephson frequencies up to several tens of GHz (Chesca et al. 2002).

Fig. 2.22 π two-junction interferometer

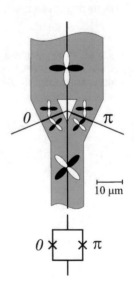

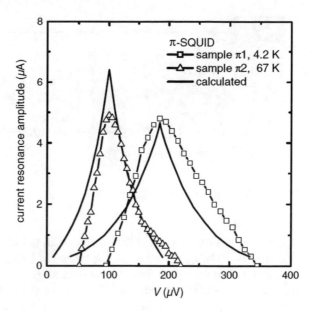

Fig. 2.23 Zero-field resonance as function of voltage in π two-junction interferometer (Chesca et al. 2002)

2.3.4 DC SQUID Characteristics

The scheme of DC SQUID presented in Fig. 2.24 contains as sensitive element double-junction interferometer. The unique sensitivity of the DC SQUIDs can be used in some situations where a variation of the quantity of interest can be transformed into variation of magnetic flux. Detail analysis of operation of DC SQUIDs presented in books (Likharev 1986; Barone and Paterno 1982) and in the review papers (Koelle et al. 1999; Fagaly 2006). The transfer coefficient H of the output voltage V on the applied flux Φ_x for DC SQUIDs is very high in comparison AC SQUIDs and can reach the value $1\,\text{mV}/\Phi_0$. For this reason amplifier nose is negligible small in comparison with intrinsic fluctuations in double junction interferometer. For DC SQUIDs with small inductance $\ell \ll 1$ (Likharev 1986; Barone and Paterno 1982), the transfer coefficient H given by expression

$$H = \frac{\omega_c I_{c1} I_{c2} \sin \phi_e}{\ell_{c+}^2 v},\qquad(2.51)$$

where $\ell_{c+} = \ell_{c1} + \ell_{c2}$ is the normalized total inductance of superconducting loop and $v = \frac{V}{V_c}$ is the normalized voltage in double-junction interferometer.

For DC SQUIDs with high inductance $\ell \gg 1$, transfer coefficient H can be written as

$$H = \frac{\omega_c \Phi_0}{2\pi L_+ I_{c1} v},\qquad(2.52)$$

where $L_+ = L_1 + L_2$ is the total inductance of superconducting ring.

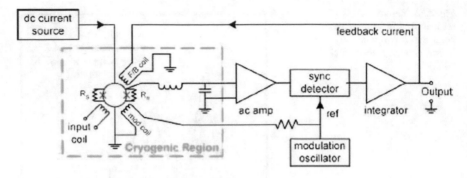

Fig. 2.24 The scheme of DC SQUID

As shown in Koelle et al. (1999), on calculating the dimensionless transfer ceof-
ficient $v_f = H\Phi_0/I_cR_N$ versus $\gamma_T\ell$ for values of ℓ ranging from 0.6 to 4 one finds
that, although v_f for a given value of $\gamma_T\ell$ decreases with ℓ, its functional dependence
on $\gamma_T\ell$ is essentially the same for all values of $\gamma_T\ell$. Thus, on normalizing curves
of v_f versus $\gamma_T\ell$ to their value at, say, $\gamma_T\ell = 1/80$, one obtains a universal curve
which is independent ℓ. Note that the value of $\gamma_T\ell = 1/80$ is an arbitrary but convenient choice suggested by the smallest value of $\gamma_T\ell = 1/80$ used in the simulations
(Fig. 2.25). The solid and dashed lines in Fig. 2.25 corresponds to different empirical
expressions for fitting (Koelle et al. 1999).

It means that energy sensitivity of DC SQUIDs is determined by the characteristics
of JJs. In the case of small thermal fluctuations $\gamma_T \ll 1$, theoretical analysis gives
result for $l \simeq 1$ energy sensitivity (Likharev 1986; Koelle et al. 1999; Kleiner et al.
2004)

$$(E_V)_{\min} = \frac{9kT}{\omega_c}. \tag{2.53}$$

More generally, in the limit $\gamma_T l < 0.2$ one finds (Likharev 1986; Koelle et al. 1999;
Kleiner et al. 2004)

Fig. 2.25 Transfer function
H of DC SQUIDs as
function of $\gamma_T\ell$ for $\beta_C = 0.5$
(Koelle et al. 1999)

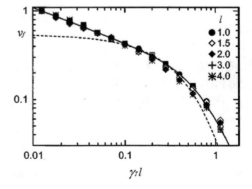

$$(E_V)_{min} = \frac{2(1+\ell)\Phi_0 kT}{I_c R_N}. \tag{2.54}$$

As followed from last expression, energy sensitivity $(E_V)_{min}$ increases with temperature, for optimized parameters, scales as $1/I_c R_N$. For LTS based high quality JJ with area $1\,\mu m^2$ and critical current density $j_c = 10^3$ A/sm^2 plasma frequency ω_p at the level 10^{12} s^{-1}. Using this data gives result for $(E_V) = 0.6 \times 10^{-33}$ J/Hz. In the study (Koelle et al. 1999) experimentally was achieved result close to this value. Energy sensitivity versus frequency for DC SQUIDs also presented in Fig. 2.19. It is clear that energy sensitivity of DC SQUIDs 10 times better than AC SQUIDs.

2.4 Superconducting Microwave Devices

In a superconductor, resistance is zero for DC current and the current flows without any dissipation. However, for AC current, superconductor shows a resistance, although the value of the resistance is very small. A phenomenological two-fluid model of superconductivity has been used to explain the general behavior of super-conductors at AC external field. Two-fluid relation for the complex conductivity (Tinkham 1996; Van Duzer and Turner 1981), lead to the expression for surface impedance of superconductor

$$Z_s = R_s + jX_s, \tag{2.55}$$

where R_s and X_s are real and imaginary parts of the impedance, respectively:

$$R_s = \frac{\mu_0^2 \omega^2 \lambda_L^3 n_n e^2 \tau}{2m}, \tag{2.56}$$

$$X_s = \omega \mu_0 \lambda_L. \tag{2.57}$$

In last equations m is the effective mass of the charge carriers, τ is the scattering time for quasiparticle, λ_L is the London penetration depth, ω is the angular frequency, and μ_0^2 is the vacuum permeability. The total carrier density is the sum of super-conducting and normal state densities $n = n_s + n_n$. The surface resistance R_s for a superconductor is proportional to the square of the frequency, whereas for a normal conductor, the surface resistance is proportional to the square root of the frequency. Both components of surface impedance of a superconductor play an important role in determining the performance of microwave devices and filters. The real part of surface resistance R_s determines the quality factor Q of the resonator, whereas the reactive component X_s determine the sensitivity to temperature variation of the wavelength of a transmission line and the long-term stability of the device.

Fig. 2.26 Surface resistance
versus frequency in YBCO
(Button et al. 1996)

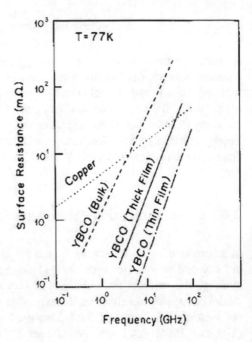

The measurement of surface resistance of superconductors is important in order to determine the suitability for its application in microwave devices. Figure 2.26 shows variation of the surface resistance with the microwave frequency for HTS YBCO bulk, thick, and thin films (Button et al. 1996). The variation of surface resistance of copper is also shown in Fig. 2.26 for comparison. The surface resistance of *YBCO* is lower than the surface resistance of copper for the frequency f^*, where f^* is the crossover frequency. The value of f^* is highest for the YBCO thin film. The surface resistance of the YBCO thin film is minimum in comparison to YBCO bulk and thick films.

2.4.1   Resonators

Superconducting resonators have been fabricated in a three-dimensional structure (cavity, dielectric resonator) and in a planar structure. Fundamental characteristics of resonators are can be determined by quality factor Q and given by an expression (Barone and Paterno 1982)

$$Q = \frac{f_0}{\Delta f},$$ (2.58)

where f_0 is the resonance frequency and Δf is the 3 dB frequency bandwidth of the resonator response. Prior to the discovery of HTS, planar structure resonators based on conventional metals had limited use due to its low Q value. The low value of the surface resistance of HTS has made it possible to realize high-Q planar resonators (Lancaster et al. 1990; Button and Alford 1992). For a *YBCO* thick-film cavity resonator operating at 5.66 GHz in *TE*011 mode, the Q value of 715,688 at 77 K has been demonstrated in Button and Alford (1992). A number of another type HTS helical resonators have been built and tested in studies (Porch et al. 1991; Peterson et al. 1989). These have been made from either thick-film or bulk polycrystalline material. A *YBCO* wire-based helical resonator has been fabricated that showed a Q of 16,000 at 77 K and an operating frequency of 355 MHz (Peterson et al. 1989).

2.4.2 Filters

One of the important components of microwave circuits are the filters. In addition to conventional applications, filters are used as a matching networks as a part of multiplexers and amplifiers. Resonators are the most common basic elements used for making filters. The main requirements for the resonators that can be used for filters are a high quality factor as well as high power handling capability. Different types of HTS based filters have been fabricated and tested in Mansour et al. (2000). These include 3D structure using dielectric resonators and 2D planar structure fabricated using HTS thin films. In study (Mansour et al. 2000) dielectric resonator–HTS based film filters have been shown to handle extremely high power levels in the range 50–100 W.

2.4.3 Antenna

Superconducting material may significantly improve the performance of an antenna or an antenna system such small antennas, matching circuits, superdirective arrays (Hansen 1990). Some of the applications were considered in study (Lancaster et al. 1992). The radiation efficiency η of a short antenna is determined by the expression

$$\eta = \frac{R_t}{R_t + R_l},\qquad(2.59)$$

where R_l is the ohmic loss resistance, R_t is the radiation resistance of antenna. Using HTS material in place of a normal conductor leads to an increase of the efficiency due to the smaller value of R_l (Khare 2003). Figure 2.27 is a schematic of HTS loop antenna, "H"-type patch (Chaloupka et al. 1991), and triangular patch antenna, along with the dipole antenna (Chung 2001). The high value of Q and small substrate height in HTS based antenna results in a narrow bandwidth between 0.85 and 1.1%. An increase of substrate height can broaden the operation bandwidth of an antenna. However, the thickness of the substrate is restricted by critical value (Khare 2003).

Fig. 2.27 Schematic
description of HTS antenna

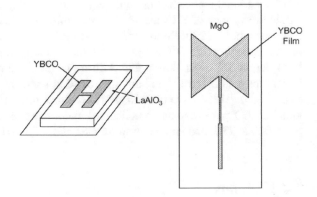

2.4.4 Delay Lines

Delay lines are useful devices in electronic warfare systems, satellite communication systems, and in realizing transversal filters for analog signal processing applications. Typical delays needed for satellite communication systems are in the range of 100–300 ns. Traditional technologies for delay lines include surface acoustic wave devices and transmission coaxial cables in the form of a coil. Superconducting delay lines offer the highest bandwidth with much lower insertion loss in a highly compact design. Several HTS planar transmission lines such as stripline, microstrip, and coplanar lines could be used to realize delay lines. Using YBCO thin film deposited on both sides of a 50-mm-diameter, 25-μm-thick LaAlO$_3$ wafer, a nondispersive delay line with 20 GHz of bandwidth has been fabricated that produces a delay of 22.5 ns with an insertion loss of 5 dB (Talisa et al. 1995).

2.5 Superconducting Devices in Meteorology

2.5.1 Voltage Standards

Physical foundation of operation voltage standard related with Shapiro steps in IV curve of externally radiated JJ (Shapiro 1963). It is well known that Shapiro steps in IV curve arises in voltages

$$V_n = n\frac{\hbar\omega}{2e}. \tag{2.60}$$

Due to that frequency can be measured with high accuracy $\delta\omega \sim 10^{-14}$, Josephson voltage standards allows to control voltage with the same accuracy. Traditional voltage standards use complex comparator schemes, due to small jumps of current at high voltages (Dziuba et al. 1974). Using integral circuit technology (Niemeyer et al. 1986) gives possibility to get serial connecting about $N \sim 1000$ of JJs and as result reaching of output voltage at the level 1 V. In this approach accuracy of voltage standard reach the value $\delta V \sim 10^{-10}$ V. A new programmable Josephson voltage standard first proposed and fabricated by Hamilton et al. (1989) ensures fast step selection and inherent stability. This type of voltage standard is based on a binary sequence of series arrays of overdamped JJs. The IV curve remains single-valued even under microwave radition. This allows the voltage step to be rapidly selected by just applying the DC bias current to the appropriate array sections. Moreover, the larger step width provides greater stability against different type of noises.

Voltage standards with accuracy at the level of fundamental constants have been realized with binary sequences of overdamped Josephson Junctions (Schulze et al. 2000). In the study Schulze et al. (2000) circuit complexities reached the level of 70,000 large area JJs (8×50 μm^2) for using in voltage standard circuits. A 10 V constant voltage was obtained under 70 GHz microwaves with power less than 1 mW. This microwave could be propagated by coupling of Josephson oscillations of all junctions which was distributed over 64 parallel microwave parts. A new circuit for JJ voltage standards was reported by Schubert et al. (2002). In these circuits coplanar strips are used for integration of the Josephson junction into a transmission line. 10 V Josephson voltage standard was realized using circuits coplanar strips design with 17,000 tunnel junctions. At frequencies around 70 to 75 GHz these standard generate reliable quantized voltage steps at 10 V with a current-step-width of about 20 μA (Schubert et al. 2002).

2.5.2 Current Standards

The small-area Josephson junctions with nano dimensions operating on the principle of Coulomb blockade of single electrons and Cooper pairs are today very important for current standard (Likharev and Zorin 1985; Nakamura et al. 1999; Zorin 2006). A deterministic current of individual charge carriers flowing through the pump under an AC drive of frequency f, can be presented as (Lotkhov et al. 2001)

$$I = ef \tag{2.61}$$

with an accuracy of 10^{-8}, can be used for the electrical current standard (Lotkhov et al. 2001). Significant improvement of the quantum standard of current can be obtained transferring Cooper pairs instead of electrons (Niskanen et al. 2003). Increase of the pumping current in single-charge devices up to the level of 1 nA can impact the quantum meteorology of electrical units. With the increase of pumping current,

Fig. 2.28 The solid state based meteorological triangle (Brake et al. 2006)

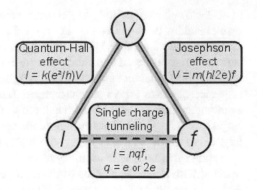

cryogenic current comparators provide accurate comparison of currents. Finally, this means completion of the meteorological triangle (Piquemal and Geneves 2000), shown in Fig. 2.28 (see also Brake et al. 2006).

2.6 Multi-terminal Devices

The multi-terminal Josephson Junctions generalizes the ordinary Josephson Junctions to the case of weak coupling between several superconducting terminals (Amin et al. 2002). Compared with two-terminal junctions, such systems have additional degrees of freedom. It means that there are corresponding set of control parameters, preset transport currents and applied magnetic fields. One of the possible realizations of a multi-terminal Josephson junction the set of superconducting terminals in which a common center connected with the microbridges. Realizations of such a coupling is the system shown in Fig. 2.29a. Another type of multiterminal JJ is based on the weak coupling of bulk superconductors through the two-dimensional normal layer (Fig. 2.29b). In four-terminal junction the non-local coupling of supercurrents is established due to the phase-dependent local Andreev reflections inside the weak link (de Bruyn 1999). The properties of the weak superconducting junction are described by the CPR (The total current in each terminal as function of the phases of the order parameter in the terminals). The CPR for the structure of Fig. 2.29a can be easily obtained by solving the GL equations close to temperatures $T \sim T_c$ and the lengths of bridges d_j much smaller then the coherence length $\xi(T)$. The supercurrent flowing into the jth terminal is determined by the phases of the superconducting order parameter ϕ_k in all the terminals (de Bruyn 1999).

$$I_j^s = \frac{\pi \Delta_0^2(T)}{4ekT_c} \sum_k \gamma_{jk} \sin(\phi_j - \phi_k), \qquad (2.62)$$

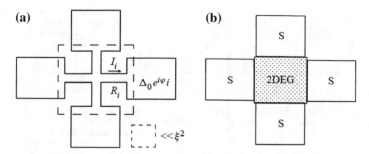

Fig. 2.29 The superconducting four-terminal JJ (**a**); the mesoscopic four-terminal JJ (weak coupling of bulk SC with two-dimensional electron gas)

where $\Delta_0^2(T)$ is the gap in the banks, γ_{jk} are the coefficients of coupling. The nonstationary regime can be qualitatively studied in the frame of the RSJ model (Likharev 1986), i.e. to the supercurrent we add the normal current $I_j^n = V_j/R_j$. The state of multiterminal JJ is described by the dynamic variables, the phases ϕ_j of the superconducting order parameter in the terminals. Let us consider the particular case of four terminals. Without loss of generality we can put $\sum_k \phi_k = 0$. Thus, it is convenient to introduce new variables for four terminal devices (de Bruyn 1998; Vleeming et al. 1999) as shown in Fig. 2.30

$$\theta = \phi_1 - \phi_2; \quad \phi = \phi_3 - \phi_4; \quad \chi = \frac{1}{2}(\phi_1 + \phi_2) - \frac{1}{2}(\phi_3 + \phi_4). \tag{2.63}$$

The four-terminal interferometer is a novel superconducting device, which is based on a four-terminal JJ. It exhibits quantum behavior by combining the properties of a single current-biased Josephson weak link and a superconducting ring interrupted by a JJ. A schematic drawing of an four terminal interferometer is given in Fig. 2.30. I is the transport current running from terminal 1 to terminal 2. The processes of switching

Fig. 2.30 The four-terminal interferometer $\phi = \phi_3 - \phi_4$, $\chi = (\phi_2 + \phi_1)/2 - (\phi_3 + \phi_4)/2$, $\theta = \phi_2 - \phi_1$

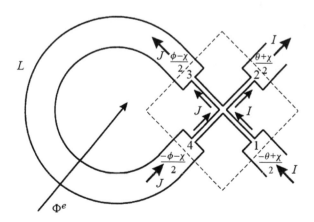

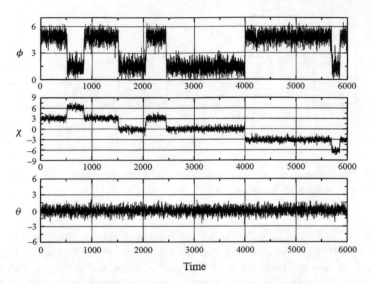

Fig. 2.31 The three phase differences ϕ, χ, θ versus time in the presence of noise (for parameters see Vleeming et al. 1999)

in the bistable state, produced by the thermal noise, were studied numerically in by solving corresponding equations and observed in experiments (Vleeming et al. 1999). Figure 2.31 display some typical results, obtained by numerical simulation. In bistable state produced transition between both flux state. Jump in ϕ are accompanied by jumps in χ by $\pm\pi$. At zero transport current no switching in θ occurs.

Chapter 3
Digital Superconductivity Electronics

Abstract The third chapter of the book is devoted to digital superconducting electronics. Firstly, brief review of this research area is presented. In second section shortly described latching Josephson logic. Foundations and basic circuits of the RSFQ logic are discussed in third section. The fourth section includes recent development in timing analysis and optimization methods of RSFQ circuit.

3.1 Brief Overview of Status of Superconducting Digital Electronics

It is well known that digital electronics related with processing and storing of binary signals. For any conventional logic device, a switching device and bit storage is the two fundamental elements for the realization of digital devices. For semiconductor digital electronics, transistor is the switch and bit storage is realized by storing charge on a MOS capacitor. Similarly, in superconducting digital electronics JJ is the switch and the bit storage is realized by storing magnetic flux quanta in an inductor realized by a superconducting loop. For practical realization of the digital circuits, the logic gates and memory elements used in the circuit should be high speed, error free and low power. Another important point for the realization of complicated logic circuits is that the fabrication process of the required devices should be reproducible and low cost. Finally, it should be possible to integrate a large number of devices into a small area (Gross and Marx 2005).

First superconducting digital element was the cryotron, which was suggested in 1950s. First cryotron was a device based on the destruction of superconductivity by a magnetic field. It consisted of a piece of wire with a winding wound over it. Current in the control winding creates a magnetic field which causes the central wire to change from its superconducting state to its normal state (Buck 1956). The major drawback of this device was the low switching speed of 10 ns limited by the L/R time constant. In addition, self-heating in the normal state increased the time for the return back to the superconducting state. Later Josephson cryotron was suggested, in which the critical current of a JJ was suppressed by a magnetic field that was typically orders of magnitude smaller than the critical field of a superconductor (Matisoo 1966).

© Springer International Publishing AG 2017

I. Askerzade et al., *Modern Aspects of Josephson Dynamics and Superconductivity Electronics*, Mathematical Engineering, DOI 10.1007/978-3-319-48433-4_3

Josephson cryotrons need much smaller control currents and smaller inductances which leads to a smaller L/R time constant. So, this result in shorter switching times.

In 1960s Josephson based logic devices research has emerged with the target of developing a very high speed and low power digital electronics technology to rival silicon integrated circuits. For digital electronics one needs two states to code the binary zeros and ones. Since tunnel JJs have two obvious states (the zero-voltage state and the voltage state, as described Chap. 2), they seemed like ideal candidates for digital electronics. A large program for developing a Josephson computer was started in IBM company in 1980s (Anacker 1980), which was stopped in 1983. IBM stopped the project to two main reasons. Firstly, Pb-alloy technology was not suitable for the fabrication of junctions acceptable spread of the critical current and Pb-alloy junctions were not suitable for repeated thermal cycling. Secondly, the IBM program was based latching logic gates based on underdamped JJs (See Sect. 3.2). Even though switching a JJ from the zero-voltage state to the voltage state is a fast process, the switch back to the zero-voltage state was a slow process. In addition, silicon integrated devices were in the meantime getting faster and JJ devices needed liquid He temperatures. So, JJ logic research was abandoned in the 1980s (Gross and Marx 2005).

Nowadays, reproducible and reliable JJ fabrication problem has been solved through the Nb-technology (Nagasawa et al. 2014; Kunert et al. 2013), and limited speed problem was solved by using the non-latching Rapid Single Flux Quantum (RSFQ) logic (Likharev et al. 1985; Likharev and Semenov 1991). In RSFQ overdamped, Josephson Junctions are used. In this non-latching logic, information is represented differently from that used in the latching voltage logic. If an overdamped junction is biased with a current slightly greater than the critical current of the JJ, the short pulses, called single flux quantum pulses, with a pulse duration being of the order of $\Phi_0/I_c R_N$ are generated. For a typical $I_c R_N$ product of 1 mV the pulse duration is about 2 ps. During a single pulse, the phase difference across the JJ evolves by 2π. One major idea of RSFQ logic circuits is the connection of JJs via inductances where the product of the JJs critical current and the inductances is constrained by an upper bound of approximately the value of the flux quantum, Φ_0. This leads to the necessity of designing very small inductances which should be provided with a high accuracy. Usually this requires sophisticated numerical procedures. Especially the parameter extraction, i.e. the calculation of designed and parasitic inductances, requires 3D numerical field computation such as Inductex (Fourie 2015) or parameter table based extraction tools such as lmeter (Bunyk and Rylov 1993). Today a large variety of complicated RSFQ circuits have been implemented (Tanaka et al. 2015; Holmes et al. 2015) and the highest speed of an SFQ device ever measured is 770 GHz (Chen et al. 1999).

In addition to the use of superconducting integrated circuits for digital electronics applications, they are also ideal partners for superconducting sensors as they have the capability of very fast digital signal processing, low power consumption and fabrication process compatibility with the sensors. These aspects of SFQ devices enable implementation of much larger number of pixels in a system (Ishida et al. 2014;

Fig. 3.1 Delay–power
relation for devices based on
various technologies (Gross
and Marx 2005)

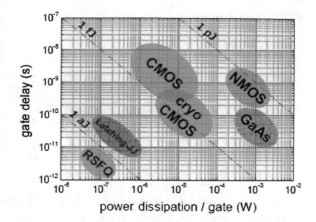

Bozbey et al. 2016). In Fig. 3.1 the delay-power characteristics of Josephson logic
devices are compared with various semiconductor logic devices. The switching delay
times of JJs are scattered below 10 ps, which is much shorter than for semiconductor
devices (Gross and Marx 2005).

The power dissipation per gate in the case using of JJ typically ranges between
0.1 and 1 μW for standard RSFQ. Static power dissipation may be reduced by an
order of magnitude if reduced static power RSFQ (Mukhanov 2011; Yamanashi et al.
2007). Even for standard RSFQ, the dissipated energy per gate cycle is typically in
the range of 10^{-18}–10^{-17} J and hence by several orders of magnitude lower than for
semiconductor devices.

Thus, the most outstanding properties of RSFQ logic is that it is possible to
develop extremely low power high speed circuits at the typical feature sizes of μm
range. As reported by the ITRS (2005) RSFQ technology is among the technologies
with lowest risk that can deliver these unique properties. However, superconductor
electronics is not intended to replace general purpose semiconductor electronics but
to provide special purpose high end solutions and digital signal processing that are
unaccessible for semiconductor electronics (Brake et al. 2006).

3.2   Latching Josephson Logic

As shown in Sect. 1.4, the switching dynamics of JJ digital circuits strongly depends
on whether the junctions are underdamped or overdamped. For the voltage state
Josephson logic underdamped junctions are required, whereas for the RSFQ logic
discussed in the next overdamped junctions are used. In this section we briefly explain
the latching Josephson logic which utilizes the switching behavior of underdamped
JJs.

Fig. 3.2 Switching of an
underdamped Josephson
junction from the "0"
($V = 0$) state to the
($V = V_g$) state"1" ($V = V_g$)
state

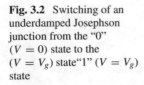

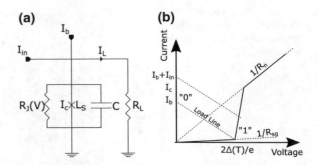

Figure 3.2a shows the circuit model for an underdamped JJ. Figure 3.2b shows the switching behaviour of an underdamped Josephson Junction from the "0" ($V = 0$) state along the load line to the "1" ($V = V_g = 2\Delta(T)/e$) state. The load line is defined by the load resistance R_L. Initially, the junction is biased with a current $I_b < I_c$. If an input current I_{in} is superposed to I_b so that $I_{in} + I_b$ is greater than I_c, the junction switches into the voltage state. The load resistor R_L should be much smaller than the sub-gap resistance R_{sg} of the junction. So that, almost all of the current through the junction is transferred to the load after the junction switches.

As explained in Chap. 1, the IV curve of such a junction is strongly hysteretic (Fig. 3.2). If the bias current is increased from $I_b < I_c$ to $I_b > I_c$ the junction switches into the resistive state and the current flows to the load resistance R_L. The junction voltage is obtained by multiplying the current I_L by the load resistor R_L. Due to its hysteretic IV characteristic the junction latches into the voltage state. In order to switch the junction back to the zero voltage state, the current has to be switched off. The junction then returns to the zero voltage state with the time constant given by t_{RC} (Likharev 1986; Barone and Paterno 1982; Gross and Marx 2005; Chen et al. 1999). However, latching Josephson logic is not suitable for ultrafast speed due to the reasons explained before. Latching Josephson logic circuits can be can be controlled by magnetically coupled gates or directly coupled gates. The reader is suggested to see the details for the latching logic circuits from (Gross and Marx 2005).

3.3   Foundation and Basic Circuits of RSFQ Logic

As mentioned in the previous section, clock frequencies of latching logic circuits are limited to a few GHz. This speed is only comparable to that of semiconductor digital circuits which do not require Helium refrigeration. Hence, the only advantage of superconducting voltage state digital circuits would be their smaller power consumption. However, this advantage is not sufficient to warrant commercial introduction of this digital technology. To overcome the limitations of the Josephson latching logic based on JJs with hysteresis, the Rapid Single Flux Quantum (RSFQ) logic has been proposed (Likharev et al. 1985; Likharev and Semenov 1991; Likharev 2001). This nonlatching logic family is based on using overdamped JJs (Fig. 3.3). Since these

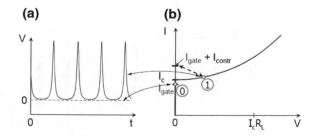

Fig. 3.3 Nonlatching logic on overdamped JJ (**a**); generation of SFQ pulses corresponding to binary codes (**b**)

junctions do no longer have hysteretic IV curve, the information can no longer be encoded in different voltage states of the junctions.

Logical "0" and "1" are represented with the existence of a single flux quantum (SFQ) pulse as shown in Fig. 3.3. The voltage pulse $V(t)$ in Fig. 3.3 is quantized and it is associated with a 2π-change of the phase across JJ. This voltage pulse has a fixed area equal to the single flux quantum Φ_0 (see Eq. (2.15)) due to the Josephson relation $V = \frac{\hbar}{2e}\dot{\phi}$. Based on this principles, SFQ pulses can be quite naturally generated, stored and transferred over a circuit. These properties enable implementation of all the basic elements of a standard logic library by using the SFQ logic.

Use of SFQ pulses for processing of digital information and analog-to-digital (A/D) conversion were developed many years ago in Clark and Baldwin (1967), Lum et al. (1977). However, complete set of RSFQ circuits was suggested by Likharev group in Moscow State University in 1985–86 year (Likharev and Semenov 1991; Likharev 2001). First experimental realization of RSFQ circuits were conducted in Kaplunenko et al. (1989), Filippenko et al. (1991) and it was achieved clock frequencies of 100 GHz. Later an alternate family of dynamic SFQ devices was suggested by Japan researcher Nakajima and coworkers (Nakajima et al. 1991; Seeber 1998).

As explained in Likharev and Semenov (1991) RSFQ logic library may consist of elementary cells such as Josephson transmission lines (JTL)s, splitters and mergers or clocked/unclocked logic gates such as AND, OR, D-Flip Flop and T-Flip Flop. In addition, for interfacing the SFQ logic to CMOS logic, SFQ/DC converters are DC/SFQ converters are implemented. For the details of these cells, please see the following sections.

As the SFQ pulse duration in the RSFQ logic is too short, binary representation is achieved as shown in Fig. 3.4. The cell is fed by SFQ pulses, which can for instance arrive from signal lines S_1, S_2 and a clock or timing line T. Each SFQ pulse coming from T input marks a boundary between two adjacent clock periods. Existence of an SFQ pulse at an input (S2 for instance) between to consecutive clock pulses is considered logic as "1" as shown in Fig. 3.4b. If there is no SFQ pulse between two consecutive clock pulses, input is considered as logic "0". In JJ circuits this representation is very natural, if SFQ pulses of the form $\int V\,dt = \Phi_0$ are used in both the S and T line. As shown above, such pulses can be readily generated and

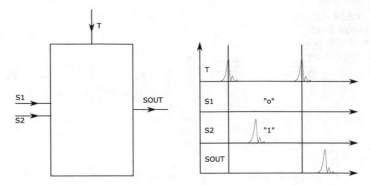

Fig. 3.4 Representation of binary units in the RSFQ logic circuits (**a**) and (**b**)

reproduced/amplified either by single overdamped JJs or by simple circuits consisting of such junctions.

3.3.1   Interfacing SFQ Pulses to Conventional Logic Circuits

SFQ logic circuits have output pulses in the order of a few ps width and 100 s of μV amplitude. Hence, it is not practical to read-out or generate SFQ pulses with conventional test equipment or interface it with the CMOS circuits. For this purpose, DC-SFQ converter circuits are used to generate SFQ pulses from CMOS logic and SFQ-DC converter circuits (Likharev and Semenov 1991) are used to generate CMOS logic outputs from SFQ pulses.

3.3.1.1   Interpretation of Inputs and Outputs of DC-SFQ and SFQ-DC Converter Circuits

As shown in Figs. 3.5 and 3.6 each rising edge of the input arriving at the DC-SFQ converter generates an SFQ pulse and each SFQ pulse arriving at the SFQ-DC converter toggles the output of the SFQ-DC circuit. These behaviors can most easily understood with the Single Data Rate (SDR) and Double Data Rate (DDR) concepts established in the CMOS logic. In SDR definition, each rising edge represents a logic "1" and in DDR definition, each rising and falling edge represents logic "1" as shown in Fig. 3.7. Basically, while interfacing the SFQ circuits to CMOS circuits, one should consider that input to SFQ circuits are represented by SDR logic while output from SFQ is represented by DDR logic. In Fig. 3.8 an example is shown by using the OR gate. If we see the input signals in between two green lines we see that Input A and Input B does not have any rising edge in between two consecutive CLK inputs (A = "0", B = "0"). Hence, the Output = "0". If we see the input signals

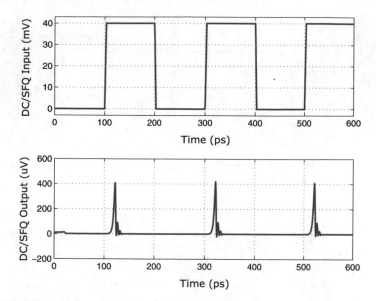

Fig. 3.5 Input–output relation of the DC-SFQ converter

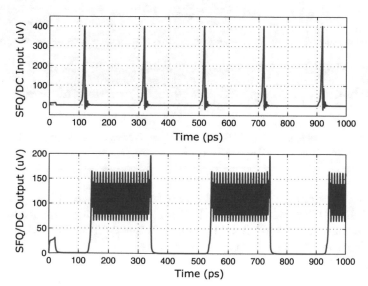

Fig. 3.6 Input–output relation of the SFQ-DC converter

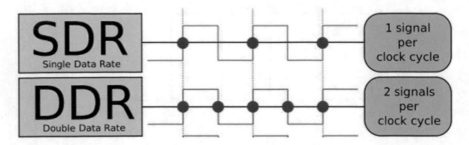

Fig. 3.7 Single Data Rate (SDR) and Double Data Rate logic representations. .s at the rising and/or falling edges represent logic "1"

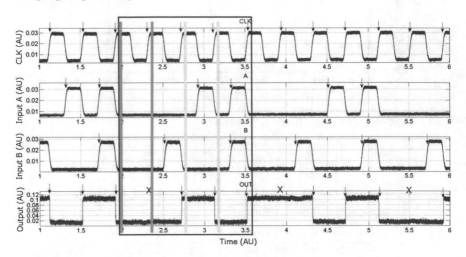

Fig. 3.8 Input–output relation of and SFQ based OR gate. *Rectangle* shows one period of the input pulse starting from A = 0, B = 0 to A = 1, B = 1

in between 2nd green line and 1st yellow line, we see that Input B has a rising edge (A = "0", B = "1"). Hence, Output = "1" (Rising edge). Similarly, inputs in between two yellow lines are A = "0", B = "1". Hence, Output = "1" (Falling edge). Arrows at the figure show the logic "1" inputs and outputs.

3.3.1.2   Layout and Fabrication

For the large scale integrated circuits composed of thousands or even ten thousands JJs a repeatable and reliable process is vital. At the moment integrated circuits are fabricated by using the $Nb/Al - AlO_x/Nb$ junctions. The reason for this is that Nb based JJs are suitable for large scale fabrication technologies and critical temperature of the material is about 9 K which is above the He based cooling systems. Some of the available RSFQ foundries are as the following:

- National Institute of Advanced Industrial Science and Technology (AIST), Japan (Nagasawa et al. 2014)
- Hypres Inc, USA (Niobium 2016)
- Fluxonics Foundry, Institute of PHotonics Technology, Germany (Kunert et al. 2013)
- MIT Lincoln Laboratory process, USA (Tolpygo et al. 2016)

Layer definitions, feature sizes and critical current densities of these foundries vary depending on the intended applications and available infrastructure. One of the most established process technology is the AIST Standard Process 2 (STP2). In Fig. 3.9 cross-section of the standard process layer definitions are shown. The process consists of four Nb layers: One ground plane and three wiring layers. A Molybdenum resistor layer below the 2nd Nb layer is used for shunting the JJs. SiO_2 insulation layers between the metal layers provide the insulation between metal layers. It is fabricated on a three-inch Si wafer. A $Nb/Al - AlO_x/Nb$ tunnel junction is used as the JJ. The minimum junction sizes and line widths are 2 μm × 2 μm and 1.5 μm respectively. The target critical current density of the JJ (J_c) is 2.5 kA/cm^2. This vertical structure is very similar to a semiconductor integrated circuit process in which the metal and insulator layers were arranged in vertically aligned stacks. There are two differences: one is the existence of JJs and the other is that the semiconductor process has a fully planarized structure (Hidaka et al. 2006).

Figure 3.10a shows the layout of JJ designed for fabrication by using AIST STP2 process. The square denoted with 2 shows the junction part and the area of this part determines the critical current of the junction. The rectangle between nodes 3 and 4 shows the shunt resistor and 5th node is the ground contact.

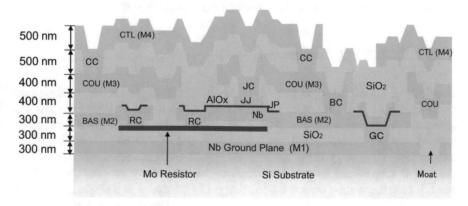

Fig. 3.9 Cross-section of a device fabricated by the standard process (Hidaka et al. 2006)

Fig. 3.10 Layout (**a**) and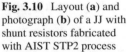
photograph (**b**) of a JJ with
shunt resistors fabricated
with AIST STP2 process

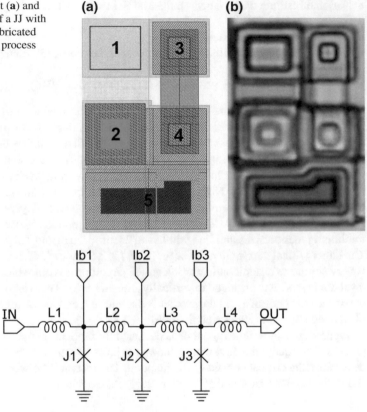

Fig. 3.11 Circuit schematics of an RSFQ Josephson transmission line (JTL)

3.3.2 Basic SFQ Circuits

In principle, almost all logic gates available in CMOS logic can be implemented
by using SFQ logic. One of the simplest and key elements of RSFQ circuits are
Josephson Transmission Lines (Fig. 3.11) that contains several JJs connected in
parallel by superconducting inductances $L = \frac{\Phi_0}{I_c}$, and DC biased to their sub-critical
current level ($I_{JTL} < I_c$). If the inductance is small enough, then a flux quantum in
the JJ-L-JJ loop cannot be stored and it propagates over the junctions. On the other
hand, if L is large, then several flux quanta could be stored in a single loop. Upon
triggering a 2π-jump of the Josephson phase in the left junction J_1 by the input signal
A, the resulting SFQ pulse developed across J_1 will cause a 2π-jump in J_2, and so
on. This behavior equivalent to a flux quantum moving from input to output. That is,
the SFQ pulse is transferred along the JTL as shown in Fig. 3.12. Similar JTL was
used for the study of sampler circuit using balanced Josephson comparator based on
overdamped JJ (Fig. 2.10, Sect. 2.2.3).

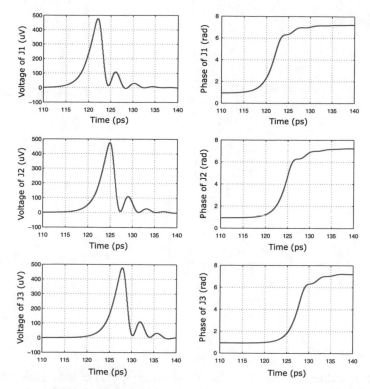

Fig. 3.12 JTL input/output (I/O) relation

This behavior can be generalized as shown in Fig. 3.13. This way, the splitting of the RSFQ pulse, i.e., reproduction of the input pulse A at each of its two outputs B and C, without noticeable decrease of the pulse voltage amplitude is achieved. One should note that JTLs and splitters transmit pulses equally well in both directions and cannot be used for isolation as shown in Fig. 3.14. In this case, critical current of J_1 should be about 1.4 times the critical currents of J_2 and J_3.

As shown in Fig. 3.13, an evident generalization of the JTL can be used to provide splitting of the RSFQ pulse, i.e., reproduction of the input pulse A at each of its two outputs B and C, without noticeable decrease of the pulse voltage amplitude. JTLs and splitters transmit pulses equally well in both directions and cannot be used for isolation as shown in Fig. 3.14. In this case, critical current of J_1 should be about 1.4 times the critical currents of J_2 and J_3.

Finally, the circuit of the buffer stage presented in Fig. 3.13. The junctions are DC biased below their critical currents. If an SFQ pulse arrives at A, it induces a 2π switching of the Josephson phase of junction J_2. This switching causes an SFQ pulse at the output. Even if the input pulse is weaker than a standard SFQ pulse, the output is a standard pulse because the circuit also provides some amplification. On the other hand, if the pulse arrives at B, junction J_1 generates a 2π pulse because of

Fig. 3.13 RSFQ splitter

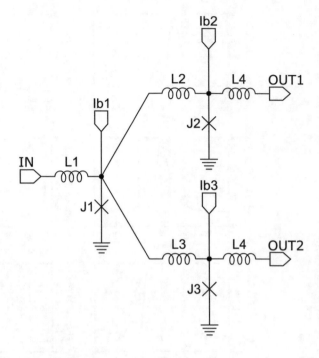

its lower I_c and no SFQ pulse is transmitted to the input A of the circuit. Different from a JTL, this circuit acts as a one-directional buffer (Fig. 3.15).

3.3.3   Simulation of RSFQ Circuits

Computer simulation is one of the most important parts of an SFQ circuit design. Conventional semiconductor circuit simulators may include JJ models that enables them to simulate Josephson circuits but as they are intended for use for semiconductor circuits, they are not very suitable for this purpose. Especially for the AC Josephson effect, the simulator has to take very small time steps to track the Josephson oscillation which results in an intensive computation (Fang and Van Duzer 1989). Josephson specific simulators such as JSIM (Fang and Van Duzer 1989) and PSCAN (Polonsky et al. 1991, 1997) are developed to make Josephson circuit simulations. In principle, these simulators are very similar to their semiconductor counterparts. In Fig. 3.16, a sample circuit schematics and its associated Jsim netlist is given. The software and user manual may be downloaded from JSIM (1989).

Even though JSIM or PSCAN is much faster than analog simulators developed for semiconductor circuits for simulating JJs, they are still much slower than hardware description language (HDL) based digital simulators. For simulating complicated circuits such as adders, arithmetic logic units etc. HDL simulators are used. In a

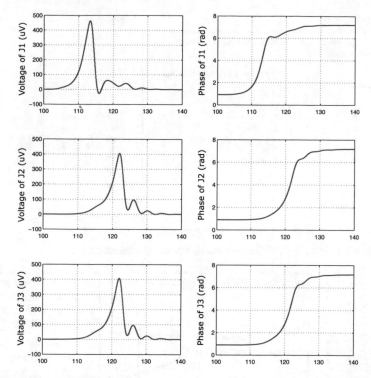

Fig. 3.14 RSFQ splitter I/O relation

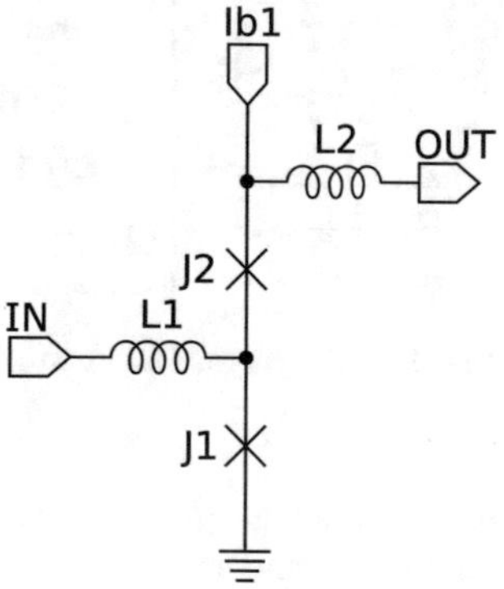

Fig. 3.15 RSFQ buffer stage

(a)   **(b)**

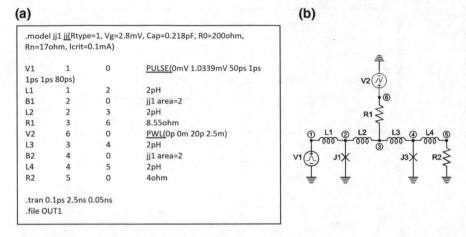

Fig. 3.16 JSIM netlist of an SFQ circuit and its associated schematics

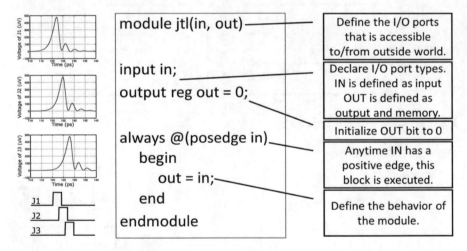

Fig. 3.17 A simple JTL circuit (Fig. 3.11) and its associated Verilog model

standard design flow, JJ level circuits are simulated and optimized by using analog simulators and their HDL models that include bias current dependent delay and critical timings are generated based on analog simulation. Then these models are used to simulate the complicated digital circuits by using HDL simulators as in the case of semiconductor circuits. In Fig. 3.17, a simple JTL circuit and its associated verilog model is presented. As HDL simulators are digital circuit simulators, SFQ pulses are modelled as digital pulses as shown in Fig. 3.17.

3.4 Timing Analysis and Optimization of RSFQ Circuits

As explained in the previous sections, for simulating RSFQ circuits, already there are some tools available such as JSIM (Fang and Van Duzer 1989) and PSCAN (Polonsky et al. 1991, 1997) for analog calculations and Verilog modules for digital simulations (Adler et al. 1997). But these simulations either take long hours to complete or give results based on only design values. Hence it is not always practical to observe the effects of timing variances that may be caused by effects such as fabrication parameter spread or the thermal noise. In this context, more specialized tools for designing and analyzing SFQ circuits are required (Fourie and Volkmann 2013). In this section we explain and model the jitter of the SFQ circuits.

3.4.1 Delay and Jitter Time in Wiring Gates

As mentioned in previous sections, the RSFQ computing has a great potential for high speed applications and it holds its position for being of relevance for future phenomena of high speed computing. Within this technology, data are obtained by using a synchronous timing scheme. In perspective of this work, it means that as the circuit frequency increases, timing errors occur more commonly. Earlier studies about RSFQ computing have showed that most of the timing errors occur in the clock distribution path and this leads the circuit to operate incorrectly (Rylyakov and Likharev 1999b; Bunyk and Zinoviev 2001; Kataeva et al. 2011; Terabe et al. 2007; Herr et al. 1999; Bunyk et al. 2003). These timing errors limit the circuits for growing larger (Rylyakov and Likharev 1999b). One other option to overcome this problem is to operate the circuit at lower frequencies to have reliable results. Many studies about timing jitter (Rylyakov and Likharev 1999b; Bunyk and Zinoviev 2001; Kataeva et al. 2011; Terabe et al. 2007) has been done and most of them pointed out that the jitter is proportional to the square root of "N" where "N" is number of JJs on a path. This approach is valid if all the logic cells would operate independently. However, since the cells are connected to each other and share the same bias line, a correlation between the neighbor cells is expected. Thus, this correlation should be taken into account when creating a model for jitter of the cascaded cells. As also pointed out in other studies (Kaplunenko and Borzenets 2001b), the delay and the jitter values are increased as the bias level reduces. The reason for this is that lowering the bias current increases the time for a shunted junction to switch into the resistive state. Furthermore, low bias level is more sensitive to noise, since the root mean square (RMS) value of the noise remains constant (Rylyakov and Likharev 1999a). As shown in Fig. 3.18, the intervals between two stable consecutive pulses for a JTL cell are about 14–15 ps, depending on the bias level.

The goal of the analysis presented in Celik and Bozbey (2013) is to identify the wiring cells used in common RSFQ libraries and to find a relation for the delay and the jitter namely for JTL, splitter and merger cells under the effects of lower power

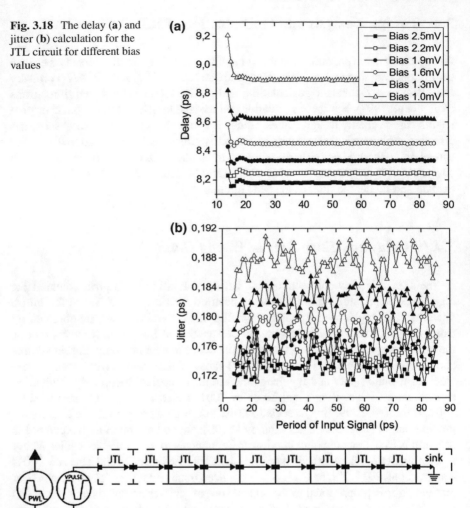

Fig. 3.18 The delay (**a**) and jitter (**b**) calculation for the JTL circuit for different bias values

Fig. 3.19 Cascaded JTL cells test bench

consumption. Also analysis of the correlation is discussed so this allows calculating of timings without doing an actual simulation. As mentioned in previous sections, JTL is one of the most basic cells which contains two JJs and used for transferring SFQ pulses along a path. Transferring SFQ pulses along a chip mostly requires a large number of JTLs to be used and it consumes quite a lot of the chip space in terms of cascaded JTLs. We start the jitter and delay analysis by using a JTL cell (Fig. 3.11) in cascade configuration as shown in Fig. 3.19. As the number of JTLs increase, the delay and the jitter values also increase accordingly.

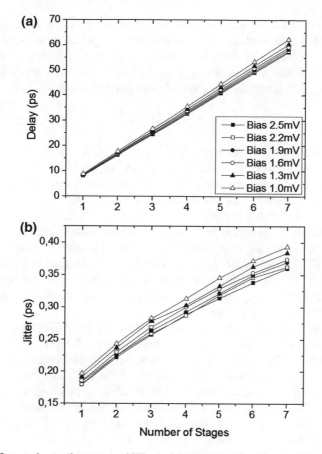

Fig. 3.20 Influence the number stages of JTL on delay time (**a**); Jitter time for JTL (**b**)

It is observed that the delay increases quite linearly with number of stages, as shown in Fig. 3.20a. As reported in Terabe et al. (2007), Kaplunenko and Borzenets (2001b), jitter is proportional with the square root of number of junctions. Calculated jitter values for the cascaded JTL cells can be seen in Fig. 3.20b. Kaplunenko and Borzenets (2001b) discuss that the jitter for cascaded logic blocks can be calculated with the first coming equation.

$$\sigma_n(x) = AI_g(x)\sqrt{n}, \tag{3.1}$$

where A is a constant that depends on temperature, frequency, and nature of the noise source, I_g is Gaussian noise spectral density and x is the distance of path that passes n JJs. With use of the equation above in order to find the fitting parameters, it is realized that Eq. (3.1) was not adequate to explain the accumulated jitter along a path. As mentioned earlier, one of our main goals of this section is to create a lookup table of the delay and the jitter for each logic gate in order to achieve fast calculation of complicated logic gates with jitter effects without using j_{sim_n}. Hereby, it is aimed to find a practical result and so it is assumed that jitter can be associated with coefficients that vary for the type of the cell and the bias level. Thus, above presented Eq. (3.1) is modified to have a fitting function for jitter.

$$\sigma = \beta\sqrt{n} + c, \tag{3.2}$$

where σ is the jitter, n is the number of junctions along the path, c is a constant that depends on the bias level and β is a constant that has its own value for the each cell. The effect of the temperature is not considered in this approach, as the most of RSFQ computing circuits operate at 4.2 K depending on today's Nb based technology. Fitting parameter "β" for the JTL gate is 9.46×10^{-14}. Fitting parameter for the different bias levels presented in Table 3.1. Similar calculations were conducted for other types of wiring cells as shown in Celik and Bozbey (2013) and fitting parameter "β" for the splitter and merger cells have been found as 13.75×10^{-14} and 17.80×10^{-14} respectively. It is found that the parameter β only depends on the cell type and does not depend on the bias value whereas the fitting parameter c is dependent on only the bias value.

Further investigation of the statistical analysis of the RSFQ logic gates show that, the timing jitter of the gates not only depends on the gate itself but also the neighboring cells. This effect is expected since the cells are connected to each other via the input and output nodes and share the same bias line. As a result, every cell combination has different delay and jitter values depending on their order on the data path. Thus, timings of each cell combinations are different and should be represented with different Gaussian functions. In usual analysis and optimization procedures (Polonsky et al. 1997; Mori et al. 2001; Intiso et al. 2005) each cell is assumed to operate independent from each other. In terms of statistics, the output timing of a cell in this situation is considered as an independent random variable. But in fact, every cell is connected to each other via the input/output nodes and the bias line, so a correlation between cells is expected. This correlation may be assessed by first calculating the timing distributions of the cells. One can find timing variations of a path by subtracting the timings of each pulse peak points of each output stages from the first input stage. Then, Gaussian representations of the stages can be obtained by using the histogram of the timing distributions. A test circuit combined with JTL, splitters (s) and mergers (m) is illustrated in Fig. 3.21. For pulse shaping, two JTL cells are used at the input side. Open ends of the circuit are terminated with sink and source cells. These cells contain no noise source and are illustrated with dashed lines throughout the article. Output timing distributions and the corresponding Gaussian functions for the test circuit are reported in Fig. 3.22.

One can see from the figure that the maxima of the Gaussian distributions are inversely proportional with standard deviation of the distributions, shown with a dashed line in Fig. 3.22, where k is a normalizing constant and σ_i is the standard deviation value of i_{th} cell. Since the timing jitter (standard deviation) of the cells is considered to be an independent random variable, it is not possible to accumulate

Bias level (mV)	Parameter c, $\times 10^{-14}$
2.5	6.38
2.2	6.52
1.9	7.17
1.6	7.32
1.3	7.97
1.0	9.58

Table 3.1 Fitting parameter "c" for the different bias levels

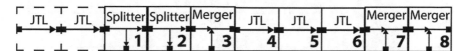

Fig. 3.21 A test circuit combined with Josephson transmission lines, mergers and splitters. Open output ends of the cells are terminated with sink cells and open input nodes of the merger cells are connected to source cells. Analyzed stages are labeled with numbers

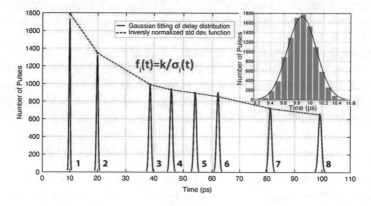

Fig. 3.22 Influence the number stages of JTL on delay time (**a**); Jitter time for JTL (**b**)

timing jitter linearly; instead variance can be used, which is the square of the standard deviation. Alternatively, the variance of a stage over a cascaded path can be calculated as

$$var = \sum_i (n_i \beta_i) + c, \tag{3.3}$$

where β_i is β parameter of i_{th} stage, which is related to the analyzed cell and the adjacent cell only as assumed. For example, β_2 is equal to β_{sm} which means that β parameter of a splitter cell followed by a merger cell combination. The number of JJs along the signal path inside the cell is represented with n and the fitting parameter which corresponds to the presence of a correlation is represented with c. The values of β and c parameters are calculated separately for each kind of cell combination used throughout the article by fitting Eq. (3.3) to the variance curves of the simulation

Table 3.2 β_i parameters for the wiring cell combinations

β parameter	Value
β_{jj}	8.03×10^{-27}
β_{jm}	6.31×10^{-27}
β_{js}	9.88×10^{-27}
β_{mj}	2.50×10^{-27}
β_{mm}	2.55×10^{-27}
β_{ms}	2.89×10^{-27}
β_{sj}	1.67×10^{-27}
β_{sm}	1.66×10^{-27}
β_{ss}	1.57×10^{-27}

results calculated by jsim Fig. 3.16 with thermal noise capability (jsim_n). The results of the β parameters are reported in Table 3.2. It is assumed that the parameter c only depends on the bias level as explained above and is calculated as 1.967×10^{-26} for 2.5 mV bias voltage.

An example calculation for finding the variance of the last stage based on Eq. (3.3) for the test cell shown in Fig. 3.21 is given in Eq. (3.4). It is just a simple linear operation for each state once the β and c parameters of the cells are known. The number of Josephson junctions on the signal path is 2 for the JTL and splitter cells, and is 3 for the merger cell (See Fig. 3.21).

$$var = 2\beta_{ss} + 2\beta_{sm} + 3\beta_{mj} + 2\beta_{jj} + 2\beta_{jm} + 3\beta_{mm} + 3\beta_{mm} + c \qquad (3.4)$$

Figure 3.23 shows the calculated and simulated variance values of the test path at each stage. Other combinations of wiring cells and logic gates are reported in Celik and Bozbey (2013) and it is shown that similar to Verilog models, a cell can be represented as a black box that has its own timing delay and timing variation, which is a Gaussian distribution with mean (μ) and standard deviation (σ). Using this model, timing distributions of the several cell combinations can be calculated with only simple linear operations. Also, output probabilities of clocked cells and the cells that are sensitive to timing variations along a path can be found. Hence, using the statistical calculations, it is possible to obtain acceptable results about the timing performance of a gate and related path instantly, rather than using time-consuming stochastic simulations.

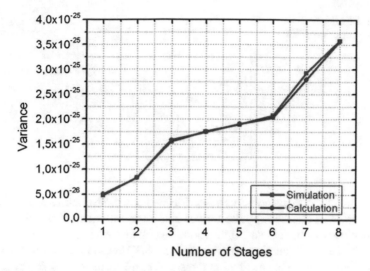

Fig. 3.23 Calculated variance values for combinations of wiring cells given in Fig. 3.21. For calculation of variance, parameters shown in Table 3.2 are used

3.4.2 Optimization of SFQ Circuits

3.4.2.1 Analog Circuit Optimization

One of the fundamental cells that provide the interface between the analog world and SFQ circuits is the 1-bit comparator circuit, which is the key element in analog to digital converter circuits and detector read-out circuits. There is not a universal optimum comparator circuit that suits for all the applications as the performance of the comparator heavily depends on the input and output circuitry. One of the main figure of merit in a comparator circuit is the gray-zone width that determines the 0-1 transition point around the threshold value where the comparator output may not reflect whether the input is above or below threshold for sure. The gray zone describes decision uncertainty of the comparator. The decision of the comparator is affected by thermal noise which causes a finite transition width between logical 0 and 1 with respect to the input signal. In this section an optimization tool to find the possible minimum gray-zone width is explained.

Quasi-one Junction SQUID (QOS) is a commonly used comparator circuit to convert analog signals into digital data in RSFQ circuits. Considering today's semiconductor technology, it has many advantages such as low power consumption, high input bandwidth, and sensitivity to the signal level. The used topology for the optimization process is given in Fig. 3.24 (Kaplunenko and Borzenets 2001a). In the QOS design, there are 5 important parameters to be optimized, which are three junction critical current values, inductance L_1, and bias current of these junctions. The current that flows through coupling coil induces current on QOS. In Fig. 3.24, G, Q

Fig. 3.24 Circuit of
quasi-one-dimension SQUID
(QOS)

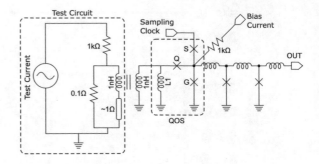

junctions, and L_1 inductance forms a interferometer and the circuit shows periodical
behavior. When a clock SFQ pulse arrives and if the current on G junction exceeds
the critical current, both G and Q junctions switch and expose an SFQ pulse to the
output. Otherwise, S junction switches and no SFQ pulse is driven to the output.
However, QOS is unstable nearby the threshold current due to the Johnson noise of
shunt resistor. The aim of this approach is to mitigate the effect of thermal noise over
decision maker close by threshold currents.

In the study Tukel et al. (2012), it was used particle swarm optimization (PSO)
algorithm for the optimization of parameters. This algorithm is inspired by the inter-
active communication of birds seeking food (Kennedy and Eberhart 1995; Poli et al.
2007; Eberhart and Kennedy 1995). In PSO, there are several particles on multidi-
mensional space searching for an optimum value of objective function and they keep
the best value they found so far. Then when they are computing the next position
they take into consideration their own best positions and the swarm's best position
called the global best. The velocity function of particles is given below:

$$V_i^{t+1} = aV_i^t + b_1(pb_i^t - x_i^t) + b_2(qb_i^t - x_i^t), \tag{3.5}$$

$$x_i^{t+1} = x_i^t + V_i^{t+1} \tag{3.6}$$

In this expression, V_i^t is velocity vector of the ith dimension on time t, pb_i^t and qb_i^t
are the particles best and global best locations of the ith dimension on time t, x_i^t is the
current position of the ith particle and a, b_1, b_2 are the constant weight parameters of
the PSO algorithm. The weight parameters are determined by both previous studies.
Additionally, some modifications have been done over PSO. For instance, if one of
the particles reaches the global best value and its velocity decreases enough, it is
pointless and waste of workload to roam around that point. Therefore, at that point,
a random set of parameters are assigned to that particle so it can look another place
for a better gray zone value as an explorer. The other modification, which has been
used, is the reflecting wall technique (Robinson and Rahmat-Samii 2004; Xu and
Rahmat-Samii 2007). If any parameter of a particle exceeds the process limitations,
that parameter is set back to the limit value and their velocity vector is multiplied by

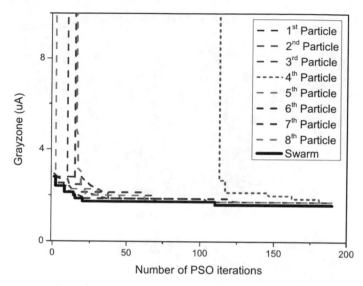

Fig. 3.25 Gray zone width of particles on PSO

minus one. Figure 3.25 shows the grey-zone evolution of each particle during each iteration.

To show the movements of particles, a two-dimensional PSO session has been done. The S junction size, the inductance, and bias current has been set to the constant values and other two values were determined by PSO. The calculated parameter sets are shown in Fig. 3.26a. As seen in Fig. 3.26a, particles search all the areas. The density is dependent to the gray zone value on that point. Particles focus on the areas which QOS operates well. Particles traces are given in Fig. 3.26b–i, respectively. When 2nd, 7th, and 8th particles were roaming around the global best point, the 4th particle was exploring other regions. Afterward, the last particle fell in the gravitational field of the global best and all particles were converged the same point. Also, it can be seen that 1st, 3rd, end 4th particles swung away from global best values several times. That is because they were sent off if their velocity vectors get closer to the zero. Thus, that modification dynamically keeps particles' searching role alive and the computational load is used more efficiently.

Table 3.3 shows 4 different optimized QOS circuit parameters. Figures 3.27 and 3.28 show the photographs and experimental results of these QOS circuits.

3.4.3 Design of RSFQ Wave Pipelined Kogge–Stone Adder

Since the invention of computers, the calculation of arithmetic and logic operations using digital circuits has been one of the leading problems in processor designs.

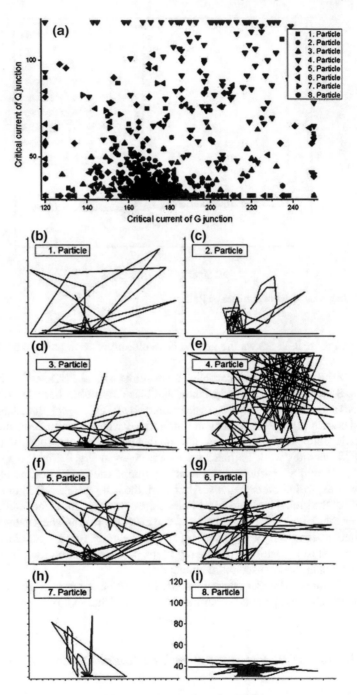

Fig. 3.26 Swarm and particles behaviors on two-dimensional spaces

Table 3.3 Fitting parameter "c" for the different bias levels

Parameter	I_{CB} (μA)	I_{CG} (μA)	I_{CS} (μA)	L_{IN} (pH)	I_B (μA)
Design 1	176	30	66	40.9	207
Design 2	233	49	66	16.12	275
Design 3	182	60	94	30.42	210
Design 4	234	79	94	12	270

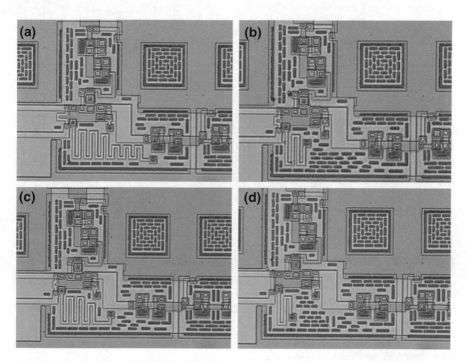

Fig. 3.27 Photographs of QOS designs 1 (**a**), 2 (**b**), 3 (**c**), and 4 (**d**)

The challenge has been to compute more operations with less clock cycles by using additional specific logic circuits. One of the most fundamental processes is addition; in which the carry bit should be transferred from the least significant bit to the most significant one. A wide range of digital circuit designs have been sustained for specialized faster addition operation. One of these adder algorithms is Kogge Stone Adder which does faster calculation with fewer levels and minimum fan-out compared to today's adders despite the only disadvantage of having an excessive amount of wiring.

In this section, a custom RSFQ based, wave pipelined, Kogge Stone Adder which was proposed in study Ozer et al. (2014) to be used later in an arithmetic logic unit (ALU). Two different design methodologies have been considered. In the first approach, it was used standard logic gates for the whole adder design. In the second

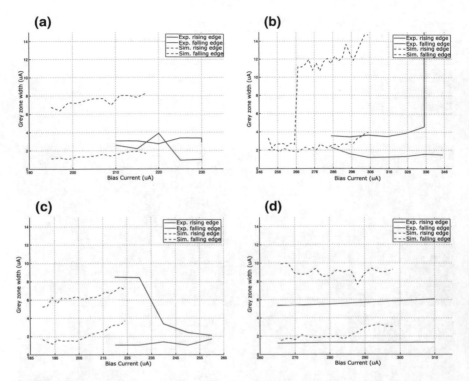

Fig. 3.28 Grey-zone widths versus bias current for QOS designs 1–4 shown in Fig. 3.27

approach, utilization to compound gate design with adjustments over component parameters is done by using PSO and statistical timing analysis tool (STAT), to increase both efficiency and bias margin (Gross and Marx 2005; Chen et al. 1999; Klein and Herrel 1978).

Today, logical functions in RSFQ require clock signal (Dorojevets et al. 2001) and because of this intrinsic feature, the maximum operation rate depends on the clock signal speed which is in direct proportion to setup and hold times of those clocked logic gates (Bunyk et al. 2001). There are also two other well-known obstacles, namely wire delays and jitter, for high speed operation at high clock rates. Since the gate and interconnect delays are low in RSFQ, propagation delay in wiring dominates especially in complex circuits even if passive transmission lines are utilized, as they need driver and receiver cells (Bunyk et al. 2001). One of the most universal digital circuits for almost any application and the ubiquitous fundamental building block for ALUs is the Adder. There are various solutions to formulate the binary addition process. Especially within the last five decades, improvements to circuit speed have been done by consuming more area in the chips for the same operation. Different architectures have been proposed and come into use as bit-serial, which operates on a single bit or digital-serial which operates on a small group of bits sequentially with a high processing speed (Dorojevets et al. 2013a). Different topologies have been designed extensively and after using many techniques, further development is almost exhausted with only minimal improvements (Dorojevets et al. 2013a).

In the paper Ozer et al. (2014), was inspected adder topologies and then determined and built an adder circuit in RSFQ logic, which is proposed to be used later in an ALU. The 8 bit adder uses wave pipeline for faster operation and clock signals are carried with data in order to minimize timing differences. As known, in most of the complicated circuits, some logic blocks are utilized a number of times. In CMOS, these blocks are optimized as compound gates such as And-Or-Invert, Or-And-Invert. In this work, had a logic block used repeatedly hence it was designed a full custom compound gate by using PSO algorithm tool (Tukel et al. 2013) explained in an earlier study to increase efficiency and bias margin of the adder circuit. Eventually, final timing adjustments are made to the whole circuit by using STAT for SFQ Cells, Celik and Bozbey (2013) to improve jitter and timing of the whole circuit.

It is useful to implement an ALU that has an instruction set including add, add-invert, and, nor, and-invert, xor and xnor. In such an ALU, one of the most complex and resource consuming arithmetic operation is addition (Filippov et al. 2011). The delay in adder circuits excessively depends on the transfer of the carry bits. The carry bit must be transferred to next significant digit in each digits' addition. Because of this reason, different implementations have been found for addition. The radix, logic depth, fan-out and wiring tracks are the main differentiating criteria for the description of adder topologies (Patil et al. 2007; Harris 2003). While carry-lookahead, conditional sum, carry select and prefix adders are attractive, most efficient topologies are the ones with the least number of logic stages with an average of two fan-out per stage and fewest wires. So the prefix adders are commonly used because of its easy expression and efficient implementations (Knowles 1999). The most efficient prefix adders are Kogge Stone Adder (KSA) (Kogge and Stone 1973) and Brent Kung Adder (Brent and Kung 1982) in terms of operation speed. While Kogge Stone Adder has many paths of similar length, Brent Kung Adder has a single long path with many small side paths (Patil et al. 2007). Kogge Stone Adder acquires minimum logical depth with recursive doubling to decrease fan-out, it also uses the idempotency property to limit the lateral logical fan-out at each node to unity (Knowles 1999). Kogge Stone Adder has very high complexity and a tremendous amount of wiring congestion. Although this also comes with a price increase in the number of lateral wires at each level, the fan-out and count of stages are very low in Kogge Stone Adder compared to Brent Kung Adder. Kogge Stone Adder topology was preferred in work (Xu and Rahmat-Samii 2007) for the Adder design because of low fan-out, which is a must in RSFQ as each gate output has a fan-out of one. In addition, less number of stages in the Kogge Stone Adder is a welcome property in the Kogge Stone Adder topology. Because of these reasons, Kogge Stone Adder topology has been chosen for a number of RSFQ groups, in which they have demonstrated successful operation (Dorojevets et al. 2013a; Filippov et al. 2011, 2012).

Generate and propagate bits are calculated at each level by using the previously linked generate and propagate bits respectively as shown in Fig. 3.29. In the last stages, all generate and propagate calculations of both current bit and previous bits have been combined to acquire the carry bit results of the addition operation. Logic

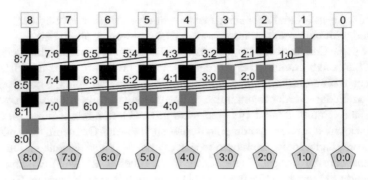

Fig. 3.29 Tree diagram of generate and propagate of bit routes

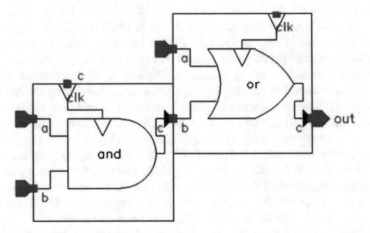

Fig. 3.30 Schematic circuit of the grey prefix cell

blocks used for the calculation of generate and propagate bits are named as Grey Prefix Cell and Black Prefix Cell. Circuit schematics of the Grey Prefix Cell and Black Prefix Cell are shown in Figs. 3.30 and 3.31.

In study Yorozu et al. (2002) designed and simulated an 8 bit wave pipelined Kogge–Stone Adder using the CONNECT cell library and have it fabricated at CRAVITY of National Institute of Advanced Industrial Science and Technology (AIST) (Hidaka et al. 2006). Kogge Stone Adder simulations are done in Verilog and correct operation has been verified. The simulation timing results are shown in Figs. 3.32 and 3.33 for input and output patterns respectively. Timing improvements were also made by using STAT as explained in the previous parts. Overall latency for the 8 bit Adder is found to be 588 ps at 2.5 mV bias voltage, where more than 37% of this latency is a result of logic gates including compound GPC. In simulations, target frequency of 25 GHz was achieved at a 2.13 mV minimum bias voltage. At the designed bias voltage of 2.5 mV, maximum working frequency of 32.26 GHz and by increasing the bias voltage to 3 mV, up to 40 GHz working frequency was achieved.

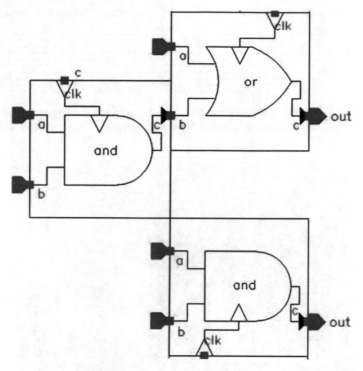

Fig. 3.31 Schematic circuit of the black prefix cell

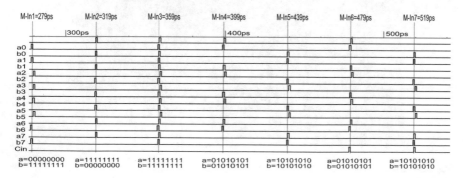

Fig. 3.32 Input pulses for Verilog Simulation. Each signal represents a digit from two 8 bit numbers. From *top* to *bottom*, first signal is the least significant bit of the first number and the second signal is the least significant bit of the second number. So inputs at 279 ps is 00000000 for first number and 11111111 for the second number

Total DC bias current, required for KSA is 0.9 A for the nominal bias voltage 2.5 mV that gives 2.25 mW of power dissipation.

Microphotograph of the fabricated Kogge Stone Adder as shown in Fig. 3.34 occupies an area of about 2.5 mm× 2.7 mm including clock splitting, DC to SFQ

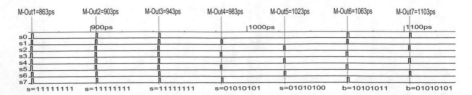

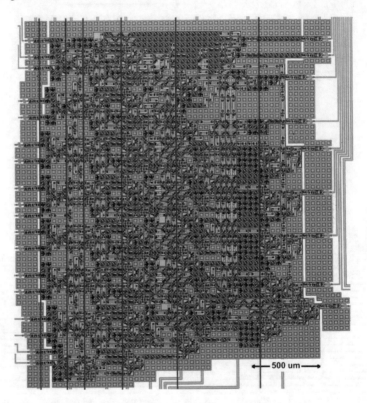

Fig. 3.33 Output pulses from Verilog Simulation. From *top* to *bottom*, each signal represents output from least significant bit to most significant bit. These pulses are consistent for adding with the inputs given in Fig. 3.32

Fig. 3.34 Microphotograph of the fabricated Kogge–Stone–Adder

and SFQ to DC converters, and it utilizes 6581 JJs in 1668 cells. The sections of the Adder shown in Fig. 3.34 from left is as follows; DC–SFQ converters, Input Signal Timing Adjustments with Clock Signal Distribution, 1st stage prefix, 2nd stage prefix, 3rd stage prefix, 4th stage result calculations and SFQ–DC converters. Prefix stages correspond to the stages shown in Figs. 3.30 and 3.31. Total number of required pads is 46. This number includes bias current inputs, signals and clock pads (Dorojevets et al. 2013b).

Chapter 4
Superconducting Quantum Bits

Abstract This chapter is devoted to describe Hamilton formalism for studying superconducting quantum bits (qubits). The first section is devoted to discuss phase, charge and flux qubits separately. For the realization of qubit operations based on JJ and their application requires the mK temperature regions. As followed from Chap. 1, anharmonic character of CPR becomes important at this temperatures and as result anharmonicity must be taken into account in consideration of JJ qubits. For this reason the influence of anharmonic CPR on qubit characteristics presented in Sect. 4.2. Description of silent qubit using anharmonic CPR given in Sect. 4.3. Flux qubit based on three JJ analyzed in Sect. 4.4. Section 4.5 is devoted to adiabatic measurements and Rabi oscillations in superconducting qubits is in Sect. 4.6.

4.1 Hamilton Formalism for Superconducting Qubits

Since last two decades, the great majority of Josephson research has focused on possible applications in the field of quantum computation (Nielsen and Chuang 2000; Valiev 2008; Wendin and Shumeiko 2007). In classical digital computation, digital electronics zero and one correspond to two distinct level of voltage. In contrast to classical computation, in quantum computation the processor takes as its input a quantum coherent superposition of ones and zeros. The quantum processor then performs a quantum mechanical operation on this input state in order to derive an output which is also a quantum coherent superposition. The basic element of a quantum computer is known as a qubit. The state of the qubit, $|\psi\rangle$ is a linear superposition of the two quantum basis states $|0\rangle$ and $|1\rangle$.

In order to analyze properties of the qubits with JJ, one has to solve the corresponding stationary Schrödinger equation with an appropriate boundary condition

$$H\Psi = E\Psi \tag{4.1}$$

where H is the Hamiltonian operator, Ψ is the wavefunction, and E is the eigenenergy. Quantum dynamics of an isolated JJ is described with the Mathieu-Bloch picture for a particle moving in a periodic potential, similar to the electronic solid state

© Springer International Publishing AG 2017

I. Askerzade et al., *Modern Aspects of Josephson Dynamics
and Superconductivity Electronics*, Mathematical Engineering,
DOI 10.1007/978-3-319-48433-4_4

theory (Likharev 1986; Zaiman 1995). In this chapter, we will describe the quantum dynamics of two type of qubits: namely phase and charge qubits. Such qubits have distinguished limiting regimes: the phase regime, $E_J \gg E_c$, is analogous to the tight-binding approximation, and the charge regime, $E_J \ll E_c$, is analogous to the near-free particle approximation (Zaiman 1995). In this chapter, we also will discuss a flux qubits, using low inductance interferometer with anharmonic CPR JJ (Amin et al. 2005; Klenov et al. 2006, 2010). As mentioned in studies (i.e., Likharev 1986; Canturk et al. 2011; Canturk and Askerzade 2011), the wave-function should satisfy the periodic boundary condition: $\Psi(\phi) = \Psi(\phi + 2\pi)$. Therefore, the required boundary condition for solving Eq. (4.1) can be given as

$$\Psi(a) = \Psi(b), \quad \Psi'(a) = \Psi'(b),$$

where $a = \phi_{\min}$ and $b = \phi_{\min} + 2\pi$ are the lower and upper bounds of phase ϕ such that a and b depend on the variation of i_b as well as α. Note that the value of a and b are different for phase and charge qubits. Additional details about a and b are given below.

The behavior of qubits will be analyzed in terms of the Hamiltonian formalism (Nielsen and Chuang 2000; Valiev 2008; Wendin and Shumeiko 2007). The general form of a Hamiltonian for a circuit containing qubits of various types is as follows

$$H = \sum_j \left\{ K(\hat{n}_j) + U(\phi_j) \right\}, \tag{4.2}$$

where $K(\hat{n}_j)$ is the kinetic energy, which is related to the electrostatic energy, and $U(\phi)$ is the Josephson energy in units $E_J = \frac{\hbar I_c}{2e}$. The Josephson energy, together with the magnetic field energy stored in the inductance of the circuit, determines the potential energy of the qubit system (see below). In the Hamiltonian formalism, the number $\hat{n}$ of Cooper pairs and the Josephson phase ϕ are conjugated operators that are related as follows

$$\hat{n} = -i \frac{\partial}{\partial \phi}. \tag{4.3}$$

For operators of number of particles $\hat{n}$ and phase ϕ is true well known quantum-mechanical uncertainty relation of the form

$$\Delta \hat{n} \Delta \phi \geq 1. \tag{4.4}$$

The CPR for JJ in general case contain all harmonics (see Sect. 1.2.1). As shown in Bauch et al. (2005), Lindstrom et al. (2006), at low temperatures it is important influence of second harmonic in $I_s(\phi)$ dependence. For this reason in this chapter potential energy in Eq. (4.2) will be taken in general case as

$$U(\phi) = E_J \left\{ \cos \phi - \frac{\alpha}{2} \cos 2\phi \right\}. \tag{4.5}$$

Fig. 4.1 Circuit model of a
phase qubit on single JJ

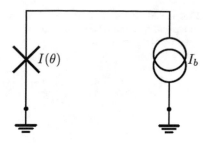

It is clear that the case of $\alpha = 0$ correspond to the JJ with harmonic CPR.

4.1.1 Phase Qubit on a Single JJ

A circuit model of a phase qubit system using single JJ is shown in Fig. 4.1. Corresponding Hamiltonian with generalized CPR can be written as Klenov et al. (2006), Klenov et al. (2010), Canturk and Askerzade (2011)

$$H = -E_c \frac{\partial^2}{\partial \phi^2} + E_J \left\{ i_b \phi + \cos \phi - \frac{\alpha}{2} \cos 2\phi \right\}, \tag{4.6}$$

where $i_b = I_b/I_c$ is the ratio of the bias current applied to the system; and ϕ denotes the phase difference. In the case of small external bias current and without anharmonicity of CPR potential energy has a form (Fig. 4.2). Energy spectrum of phase qubit on single JJ with harmonic CPR, which coincides with the spectrum of harmonic oscillator (see for example Landau and Lifshitz 1965)

$$E_{n0} = \hbar \Omega_p (n_0 + 1/2), \tag{4.7}$$

where is the plasma frequency of $\Omega_p = \omega_J (1 - I_b/I_c)^{1/4}$ (Likharev 1986; Barone and Paterno 1982), which correspond to phase oscillations near minimum potential

Fig. 4.2 Potential energy
profile of phase qubit on
single JJ

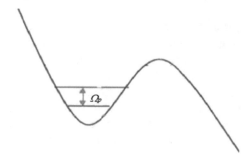

Fig. 4.3 Microphotograph of a phase qubit based on single JJ (Martinis et al. 2002)

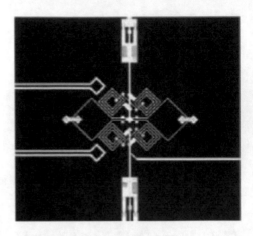

energy. The Josephson frequency ω_J is determined by the expression

$$\omega_J = I_c/2e. \tag{4.8}$$

As followed from last expression, distance between 0 and 1 energy levels is controlled by external current I_b.

Phase qubit based on single JJ firstly was experimentally realized in Martinis et al. (2002). Microphotograph of phase qubit presented in Fig. 4.3. The ratio of Josephson energy to Coulomb energy in fabrication of this qubit scheme E_J/E_C is about 10^4.

4.1.2 Charge Qubit on Single JJ

A circuit model of a charge qubit system using single JJ is shown in Fig. 4.4. The Hamiltonian of the charge qubit system (Wendin and Shumeiko 2007; Canturk et al. 2011), associated with generalized CPR can be written as

$$H = E_C(\hat{n} - n_g)^2 - E_J \left\{ i_b\phi + \cos\phi - \frac{\alpha}{2}\cos 2\phi \right\}, \tag{4.9}$$

where $E_C = Q_g^2/(2C_\Sigma)$ is the electrostatic energy (Cooper pair charge energy) depending on gate voltage V_g and the capacitor $C_\Sigma = C_g + C_J$. In addition, $\hat{n} = -i\frac{\partial}{\partial\phi}$ is the dimensionless 'momentum' operator that refers to the number of Cooper-pair on the island and has a physical meaning of charge Q accumulated on the junction capacitor C_J in the units of double electronic charges (i.e., $\hat{Q} = 2e\hat{n}$). Furthermore, $n_g = C_g V_g/(2e)$ refers to the dimensionless charge number used to externally control the system (Nakamura et al. 1999). Figure 4.4 illustrates a single Cooper pair box for a charge qubit including a gate voltage V_g and a gate capacitance C_g.

Fig. 4.4 Circuit model of a
charge qubit using single JJ

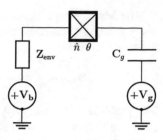

With the help of washboard potential in Eq. (4.5) for non-zero bias current as well
as α, the upper and lower bound of the interval of periodicity can be determined from
$U(\phi)$ which has a stable minimum ϕ_{min} determined from

$$f(\phi) = \frac{dU}{d\phi} = -i_b + \sin(\phi) - \alpha\sin(2\phi) = 0$$

using a root finding technique. In this case, the interval for periodic boundary will be
$a = \phi_{min}$ and $b = \phi_{min} + 2\pi$. Periodical solutions in the case of non-zero bias current
were discussed in detail (Likharev 1986 and Canturk et al. 2011) using wavepacket
approach.

After substituting Eq. (4.9) into Schrödinger equation (4.1), we can obtain a Mathieu type eigenvalue equation:

$$\frac{d^2\Psi}{d\phi^2} - p\frac{d\Psi}{d\phi} + q\Psi = -\varepsilon\Psi, \tag{4.10}$$

where $\varepsilon = E/E_C$, and the terms $p = 2in_g$ and

$$q(\phi) = \frac{E_J}{E_C}\left[i_b\phi + \cos\phi - \frac{\alpha}{2}\cos 2\phi\right] - \frac{|p|^2}{4}.$$

Here, we will present the evaluation of expectation value of the supercurrent operator
given in equation $(\hat{I}_s/I_c = \sin\phi - \alpha\sin 2\phi)$ and the number operator $\hat{n} = -i\frac{\partial}{\partial\phi}$
within the interval $[a, b]$ in the lowest band for the charge qubit.

First of all, the expectation value of supercurrent (Likharev and Zorin 1985) can
be obtained from

$$i_s = \langle\Psi|\hat{I}_s/I_c|\Psi\rangle = \frac{1}{b-a}\int_a^b \Psi^*\{\sin\phi - \alpha\sin 2\phi\}\Psi\, d\phi. \tag{4.11}$$

Similarly, the expectation value of number operator $\hat{n}$ can be determined as

$$\langle\hat{n}\rangle = \langle\Psi|\hat{n}|\Psi\rangle = \frac{1}{b-a}\int_a^b \Im\left\{\Psi^*\frac{\partial\Psi}{\partial\phi}\right\} d\phi. \tag{4.12}$$

The Mathieu type eigenvalue problems defined in Eqs. (4.1) and (4.12) can be discretized using the finite difference approach discussed in Canturk et al. (2011) on a discrete lattice:

$$e\Psi_{j-1} + f_j\Psi_j + e^*\Psi_{j+1} = \lambda\Psi_j, \forall j = 0, 1, 2, \ldots, N - 1.$$

The problem is that we want to get rid of the upper and lower end wavefunctions ($\Psi_{-1} = \Psi_{N-1}$ and $\Psi_N = \Psi_0$) in the form of boundary conditions given in Eq. (4.2). After that, the mathematical discretization of these equations approximates the continuum behavior of the system to obtain the eigenvalue problem of the form:

$$\mathbf{A}_0\Psi_n = \lambda_n\Psi_n, \quad n = 0, 1, 2, \ldots \tag{4.13}$$

where Ψ_n is an eigenfunction associated with eigenvalue λ_n for an arbitrary eigenstate n; and $\mathbf{A}_0$ is a complex $N \times N$ periodic tridiagonal coefficient matrix

$$\mathbf{A}_0 = \begin{pmatrix} f_0 & e^* & 0 & 0 & 0 & \cdots & 0 & e \\ e & f_1 & e^* & 0 & 0 & \cdots & 0 & 0 \\ 0 & e & f_2 & e^* & 0 & \cdots & 0 & 0 \\ 0 & 0 & e & f_3 & e^* & \cdots & 0 & 0 \\ 0 & 0 & 0 & e & f_4 & \cdots & 0 & 0 \\ \vdots & \vdots & \vdots & \vdots & \vdots & \ddots & \vdots & \vdots \\ 0 & 0 & 0 & 0 & 0 & \cdots & f_{N-2} & e^* \\ e^* & 0 & 0 & 0 & 0 & \cdots & e & f_{N-1} \end{pmatrix}. \tag{4.14}$$

The coefficients of the periodic tridiagonal matrix in Eq. (4.14) are given as

$$e = (1 + hp/2), \quad f_j = (h^2q_j - 2), \quad \lambda = -h^2\varepsilon, \\ h = \frac{b-a}{N+1}, \tag{4.15}$$

where e^* is the complex conjugate and h is the step size for discretization scheme.

The numerical solution of Eq. (4.10) is calculated using the conventional LAPACK eigenvalue solver. Note that the coefficient matrix in Eq. (4.15) requires an $N + 2$ data storage space in its present form. In the literature Bjorck and Golub (1977) and Evans and Okolie (1982), the linear system solution with a real symmetric periodic tridiagonal coefficient matrix is widely studied, but the eigenvalue problem (Canturk et al. 2011) formulated in the present section requires an efficient design and stable algorithm for finding the eigenvalues of such a matrix. The number of subintervals is preferred to be $N = 3000$ with sufficient accuracy, the reason for which the preference will be explained. Energy spectrum of JJ based charge qubit without Josephson coupling term presented in Fig. 4.5. Inclusion of Josephson coupling leads splitting of spectrum and appearance of band structure (Fig. 4.6).

Fig. 4.5 Energy spectrum of charge qubit without Josephson coupling

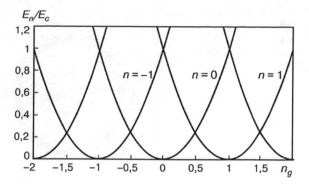

Fig. 4.6 Energy spectrum in the case of Josephson coupling (band structure of spectrum)

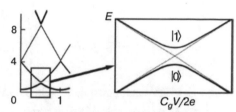

Fig. 4.7 Scanning-electron micrograph of charge qubit on JJ (Nakamura et al. 1999)

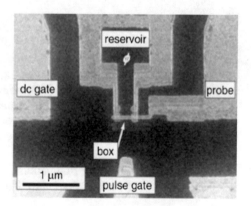

Charge qubit based on single JJ firstly was realized in study Nakamura et al. (1999). Scanning-electron micrograph of charge qubit presented in Fig. 4.7. The ratio of Josephson energy to Coulomb energy E_J/E_C in this study was taken is about 0.3.

4.1.3 Flux Qubit on Single JJ Interferometer

For the flux qubit based on a superconducting interferometer with a single JJ (Fig. 4.8a), the Hamiltonian acquires the following, more particular form

Fig. 4.8 Single junction flux qubit: equivalent circuit diagram **a** symmetric double-well potential

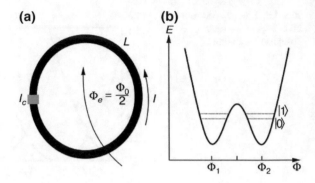

$$H = E_C \hat{n}^2 - E_J \left\{ -\cos\phi + \frac{(\phi - \phi_e)^2}{2\ell} \right\}, \tag{4.16}$$

where $E_J \frac{(\phi - \phi_e)^2}{2\ell}$ is the magnetic field energy stored in the superconducting circuit, $\ell = \frac{2\pi L i_c}{\Phi_0}$ is the normalized inductance of this circuit, and C is the capacitance of the junction.

The potential $U(\phi)$ of the phase qubit for $\phi_e = \pi$ (i.e., for $\Phi_e = \Phi_0/2$) has a double well shape (Fig. 4.8b). With neglect of quantum tunneling between the wells, the energy levels in these wells are identical and, hence, the ground state is doubly degenerate. With allowance for the tunneling, this state splits to form a real two level system, which is a very important circumstance for the phase qubit. The splitting of the ground energy level $E = E_0 \pm \Delta E$ in this quantum mechanical system provides a basis for the qubit under consideration. The distance between the two sublevels is determined by the tunneling length, which is much smaller than the distance between levels in a single well potential.

In the small inductance approximation ($\ell \ll 1$), the Josephson energy $-E_J \cos\phi$ is a perturbation in the zero order Schrödinger equation

$$\left(\frac{Q^2}{2C} + E_j \frac{\phi^2}{2\ell} \right) \Psi = E_n \Psi. \tag{4.17}$$

As can be seen from Eq. (4.17) coincides with the equation of a quantum mechanical oscillator with the frequency $\omega^2 = \frac{1}{LC}$, which possesses the spectrum of quantum oscillator given by Eq. (4.7). The splitting ΔE is proportional to E_J and can be determined as a correction to the energy spectrum (4.7) in the first order of perturbation theory, which leads to the following expression (Askerzade and Amrahov 2010)

$$\Delta E_n = E_J \left\{ 1 - \cos\phi_e \exp\left(-\frac{\pi^2 \hbar \omega L}{\Phi_0^2} \right) L_n \left(\frac{\pi^2 \hbar \omega L}{\Phi_0^2} \right) \right\}, \tag{4.18}$$

where $L_n(x)$ are the nth order Laguerre polynomials. In the case of a phase qubit, we are interested in the correction to the ground state level of the oscillator ($n = 0$, $L_0 = 1$) and eventually obtain the following expression for this level

$$E = E_0 + \Delta E = \frac{\hbar\omega}{2} + E_J \left\{ 1 - \cos\phi_e \exp\left(-\frac{L}{L_F}\right) \right\}. \qquad (4.19)$$

where $L_F = (\Phi_0/\pi)^2 \frac{1}{\hbar\omega}$ is the quantum fluctuation inductance that was introduced in Askerzade (2005a).

The results of the above analysis are illustrated in Fig. 4.9, which shows that the energy splitting ΔE of the ground state depends on the relative inductance L/L_F and the parameter $\cos\phi_e$ determined by an external magnetic field. For very large inductances, $L/L_F \gg 1$, the superconducting current is suppressed by quantum fluctuations and the splitting weakly depends on the inductance of the circuit. In addition, ΔE depends on the magnetic field applied to the circuit. The correction ΔE is at minimum for $\phi = 0$ and increases with ϕ_e (for the same geometry of the superconducting interferometer). The effect of the external magnetic field on ΔE is also demonstrated in Fig. 4.9.

Equation (4.19) describes the sensitivity of the energy splitting ΔE with respect to various physical parameters, which provides a possibility for the control. These parameters are important for determining the field of possible applications of the single junction qubit, solving the problem of data readout from the qubit, and preserving its coherence (Wendin and Shumeiko 2007). The ΔE value enters directly into the Hamiltonian that controls evolution of the qubit state. This circumstance makes it possible to create qubits specially intended for accomplishing particular logical operations under the action of applied magnetic fluxes of certain durations.

Fig. 4.9 Splitting of ground state energy of a flux qubit versus inductance of superconducting ring for different external magnetic flux (External magnetic flux ϕ_e changes as $\pi/4$, $\pi/2$, $2\pi/3$, $3\pi/4$ and π from *bottom to top*.)

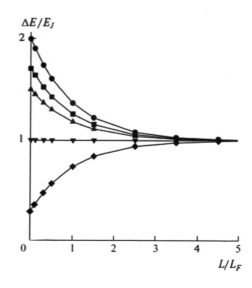

If the interferometer has a large inductance, the Josephson potential in Hamiltonian (4.16) predominates and this approximation corresponds to the case of charge or phase qubit based on single JJ. Flux qubit based on single JJ interferometer firstly was realized in study Friedman et al. (2000).

4.2 Influence of Anharmonic CPR on Qubit Characteristics

4.2.1 Phase Qubit

The presence of second harmonic in CPR leads to hump like shape in potential energy which seems to be very important for the manipulation of phase qubit. Figure 4.10 illustrates the influence of second harmonics on the potential $U(\phi)$ for various values of α. For instance, the potential has a single minimum at $\phi = 0$ for $\alpha \leq 0.5$. Whereas, it has double minima for $\alpha > 0.5$ in the ranges $-\pi < \phi < 0$ and $0 < \phi < \pi$ respectively. The authors in Goldobin et al. (2007) and Yamashita et al. (2006a) also discussed how $U(\phi)$ changes from single potential well to double potential well in the case π junctions (i.e., the junctions with negative critical current).

After substituting Eq. (4.6) into the Schrodinger equation (4.1), we can obtain Mathieu eigenvalue equation for zero bias current case:

$$\frac{d^2\Psi}{d\phi^2} + \frac{E_J}{E_C}\left\{\cos\phi - \frac{\alpha}{2}\cos 2\phi\right\}\Psi = -\varepsilon\Psi, \tag{4.20}$$

where $\varepsilon = E/E_C$ and it is called the Mathieu eigenvalue equation for describing the properties of phase qubit under the periodic boundary condition with lower bound $a = 0$ and upper bound $b = 2\pi$.

Fig. 4.10 Potential energy of anharmonic JJ

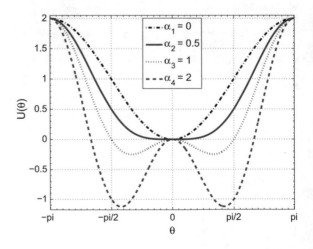

In study Klenov et al. (2006) only asymptotic solutions of Mathieu equations (4.20) were presented. Here we performed full numerical analysis of Mathieu equation (4.20) with second harmonics in phase qubit regime (i.e., $E_J/E_C \gg 1$). As followed from the results, all energy levels split into two sub-levels $\varepsilon_i^{\pm} = \varepsilon_i \pm \Delta_i$. Energy spectrum of Mathieu equations for $i = 0, 1$ using numerical calculations are presented in Fig. 4.11 coincides with the results in Klenov et al. (2006). The ground ($i = 0$) and first ($i = 1$) states of the energy spectrum are obtained and it is shown that, splitting in the ground and first excited states depend on the anharmonicity parameter α. For high values of energy scale $E_J/2E_C$, the splitting between $\alpha = 0$ and $\alpha \neq 0$ cases become large. On the other hand, it can be seen from the calculations,

Fig. 4.11 Splitting parameters versus anharmonicity α of the CPR. **a** Splitting of the ground state. **b** Splitting of the first excited state

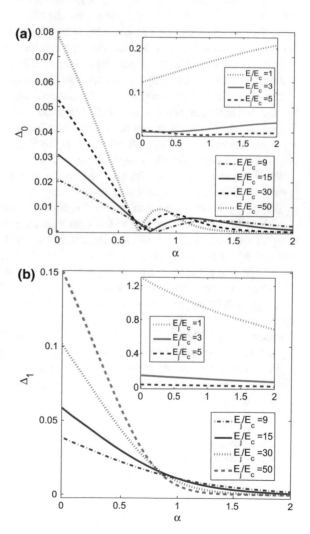

the change in splitting of ground state is smaller than the change of the first level. This means that, the first state is more sensitive to anharmonicity parameter α.

Furthermore, the numerical modeling is conducted to analyze the influence of the control parameters on the splitting of energy states $\Delta_i = \varepsilon_i^+ - \varepsilon_i^-$. Figure 4.11 presents the behavior of the splitting of energy states Δ_i for various energy scale E_J/E_C. The results for Δ_0 are presented in Fig. 4.11a within the range $0 \leq \alpha \leq 2$. As mentioned before, the authors in Klenov et al. (2006), Askerzade (2011) analytically found similar results for two-junction interferometer from an oscillatory model. Unlike our results, their findings are limited to the range $0.5 \leq \alpha \leq 1.5$ for Δ_0. In contrast to Klenov et al. (2006), Askerzade (2011), we observed fine structure in dependence $\Delta_0(\alpha)$ for different E_J/E_C values. For small anharmonicity parameter (i.e., $\alpha < 0.65$), the splitting parameter Δ_i decreases linearly with an increase in α. The results for Δ_i are in good agreement with findings of solid state theory. By fixing the amplitude of first harmonic, negative sign of second harmonic leads to an approximate linear decreasing of $\Delta_0(\alpha) \approx E_J(1 - \alpha)$ (Zaiman 1995). Similar behavior of linear decreasing in Δ_0 was obtained in our numerical results presented in Fig. 4.11a. However, compared to approximate result, the vanishing point of Δ_0 has located in the range $0.6 < \alpha < 0.9$ for various energy scales. The reason for this is that numerical results are more precise than the results obtained from preceding approximate expression. In addition, $U(\phi, \alpha)$ in Fig. 4.10 has a single minimum for $\alpha \leq 0.5$ while it has double minima for $\alpha > 0.5$.

As discussed before for Fig. 4.10, the shape of the potential was switched from a single well to a double well structure for $\alpha > 0.5$. For higher energy scale such as $E_J/E_C \geq 9$, the behavior of $\Delta_0(\alpha)$ illustrates different tendencies. For instance, $\Delta_0(\alpha)$ keeps decreasing from $\alpha = 0.5$ to $\alpha = \alpha_{\text{critical}}$ until it vanishes. The value of α_{critical} are determined from our calculation (see Fig. 4.11a) as $0.6231, 0.6007, 0.5784$, and 0.5634 for energy scales $9, 15, 30$, and 50 respectively. The height of the hump of the double well is not so high in this region and the energy levels are strongly coupled. Such behavior corresponds to two level crossing. On the other hand, for $\alpha_{\text{critical}} < \alpha \leq \alpha_{\text{max}}$ where α_{max} is given in Table 4.1, $\Delta_0(\alpha)$ has an increasing tendency as shown in Fig. 4.11a. For $\alpha > \alpha_{\text{max}}$ the hump of the double well increases so does the energy level become weakly coupled. Consequently, the second harmonics in Eq. (4.6) becomes dominant and leads to a two-level crossing.

For large values of anharmonicity parameter (i.e., $0.65 < \alpha$), we obtain results similar to Klenov et al. (2006). The maximum value of $\Delta_0(\alpha)$ peaks at different energy scales are given in Table 4.1

Table 4.1 Influence of E_J/E_C on peak position

E_J/E_C	$\alpha_{\max}$	$\Delta_{0\max}$
9	1.350	0.0045
15	1.125	0.0054
30	0.950	0.0072
50	0.875	0.0090

As also shown in Fig. 4.11, the peak position of Δ_0 depends on the energy scale E_J/E_C. Besides, the width of the peak grows with decreasing energy scale E_J/E_C in phase qubit regime. With decreasing E_J/E_C, the peak value of $\Delta_0(\alpha)$ is also suppressed. As shown in the inset of Fig. 4.11a, $\Delta_0(\alpha)$ reveals only growing tendency when $E_J/E_C < 3$ for charge qubit regime. Figure 4.11b illustrates similar results for Δ_1. As can be seen from this figure, $\Delta_1(\alpha)$ reveals monotonic decreasing behavior with an increase in the anharmonicity parameter α for all E_J/E_C.

The influence of the energy scale E_J/E_C on splitting of energy state Δ_i for fixed value of anharmonicity parameter α is presented in Fig. 4.12. This plot clearly illustrates an upward trend of Δ_i for $E_J/E_C > 50$. The reason for restricting E_J/E_C

Fig. 4.12 Splitting versus energy scale. **a** Splitting of the ground state. (*Inset*) Behavior of Δ_0 for the range $0 \leq E_J/E_C \leq 4$. **b** Splitting of the first excited state. (*Inset*) Behavior of Δ_1 for the range $0 \leq E_J/E_C \leq 4$

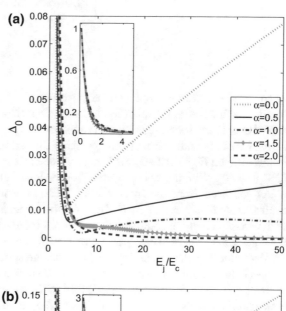

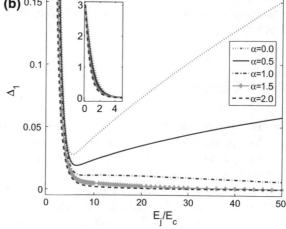

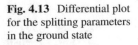

Fig. 4.13 Differential plot for the splitting parameters in the ground state

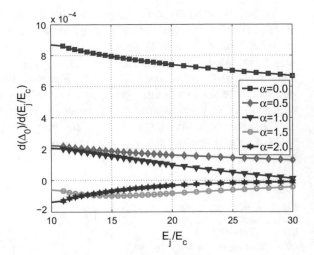

up to 50 is related to technological achievement in the realization of JJ with a very small capacitances (at a level of femto Farad (*fF*)) (Nakamura et al. 1999). The splitting of energy in ground state Δ_0 increases for energy scale $E_J/E_C > 3$. The partial derivative of Δ_0 with respect to energy scale for various α is plotted in Fig. 4.13. As followed from Figs. 4.12 and 4.13, high slope corresponds to the case of harmonic CPR ($\alpha = 0$). With increasing α up to 2, the slope falls down to zero.

As followed from the inset of Fig. 4.12a, at small $E_J/E_C < 3$, Δ_0 sharply decreases with an increase in E_J/E_C. Another peculiarity of this region is related with non-sensitivity of results to changing of amplitude of second harmonic α. The influence of the energy scale parameter E_J/E_C on splitting or energy state Δ_1 for fixed α value is presented in Fig. 4.12b. Thus, the splitting of first state $\Delta_1(E_J/E_C)$ reveals behavior similar to the case $\Delta_0(E_J/E_C)$. The inset of Fig. 4.12b shows that at small $E_J/E_C < 3$, Δ_1 decreases sharply with an increase in E_J/E_C similar to Δ_0. Note that the results for differential plots of the first level also resembles the results shown in Fig. 4.13.

4.2.2 Charge Qubits

As mentioned before, the energy scale $E_J/E_C < 1$ corresponds to the single Cooper box (SCB) charge qubit limit. In this limit, energy spectrum can be described as quasicharge approach (Likharev 1986; Canturk et al. 2011) similar to quasi-momentum representation in solid-state theory (Zaiman 1995). Energy gap Δ_0 versus anharmonicity parameter α presented in Fig. 4.14a resembles to phase qubit case in Fig. 4.11. Similar to Δ_0 which is the difference between ε_1 and ε_0 at $n_g = 0.5$, the "secondary energy gap" Δ_1 is the difference between ε_2 and ε_1 at $n_g = 1$. The

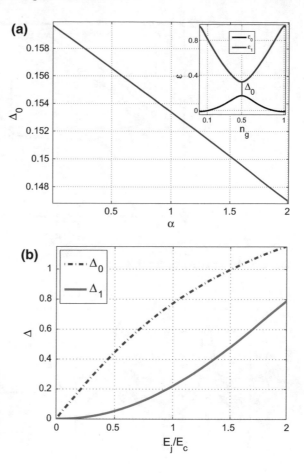

Fig. 4.14 Results for the charge qubit obtained at $n_g = 0.5$

detailed description is left to Canturk et al. (2011). Notice that Δ_i refers to energy gap in charge qubit whereas it refers to splitting of energy states in phase qubit. In Fig. 4.14b, it is illustrated the dependency of gap parameters Δ_i on energy scale E_J/E_C. This result is also qualitatively in good agreement with Fig. 4.12. However, in case of SCB the growing Δ_i with E_J/E_C has revealed a nonlinear behavior. The expectation value of number operator $\langle \hat{n} \rangle$ given in Eq. (4.13) is plotted with respect to gate number n_g in Fig. 4.15 at small bias current $i_b = 0.1$. The $\langle \hat{n} \rangle$ versus n_g is experimentally observed for SCB in study Bouchiat et al. (1999) for junction parameters $E_c = 0.215$ meV and $E_J/E_c = 0.16$. As followed from Fig. 4.15 $\langle \hat{n} \rangle$ versus n_g is not sensitive to α. The expectation value of supercurrent $i_s = \langle \hat{I}/I_c \rangle$ versus n_g is illustrated in Fig. 4.16 for different anharmonicity parameter α. The positions of the peaks in i_s versus n_g relation is the same as the peaks for $\alpha = 0$ in Canturk et al. (2011). Notice that the supercurrent is equal to zero when the bias current i_b is set to zero. The peaks at half-integer n_g values correspond to the tunneling of Cooper pairs from one electrode to another. When we compare Fig. 4.16a, b, the increasing of i_b

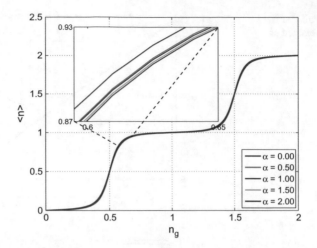

Fig. 4.15 $\langle n \rangle$ versus n_g for $i_b = 0.1$

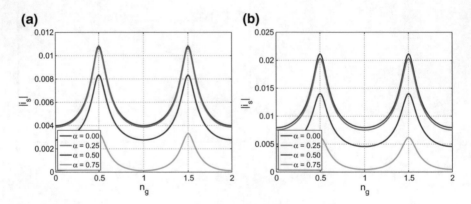

Fig. 4.16 Expectation value of supercurrent $\langle I/I_c \rangle$. **a** $i_b = 0.05$. **b** $i_b = 0.1$

leads to an increase in the magnitude of i_s while the increasing of anharmonicity parameter α leads to a decrease in the magnitude of i_s.

It was shown in Zubkov et al. (1981) that the anharmonicity of the CPR is equivalent to the introduction of an effective inductance connected in series to the JJ. The value of effective inductance is proportional to the magnitude of the anharmonicity parameter. This implies that such additional inductance leads to an increase in the impedance of the circuit for charge qubt. For that reason, the amplitude of the expectation value of supercurrent i_s in Fig. 4.16 is suppressed with increasing α.

4.3 Silent Qubit

The silent qubit is just a low-inductive two-junction interferometer using JJ with anharmonic CPR. In silent qubit, it is not required to apply half flux quantum. Such a qubit was called "silent" because of both the high protection against external magnetic field impact and the absence of any state-dependent spontaneous circular currents. The potential energy of the quantum-mechanical system described in the following papers Kornev et al. (2006), Amin et al. (2005), Klenov et al. (2010). As shown in the previous section, presence of second harmonic in CPR of JJ leads to two-hump potential (Fig. 4.17)

$$U(\phi, \phi_+) = -\frac{\Phi_0 I_{c1}}{2\pi}\left\{\cos\left(\frac{\phi}{2} + \phi_+\right) - \frac{\alpha_1}{2}\cos(\phi + 2\phi_+)\right\} \qquad (4.21)$$
$$-\frac{\Phi_0 I_{c2}}{2\pi}\left\{\cos\left(\frac{\phi}{2} - \phi_+\right) - \frac{\alpha_2}{2}\cos(\phi - 2\phi_+)\right\},$$

where $\phi = \phi_1 - \phi_2$ is the difference of the JJ phases, $\phi_+ = \frac{\phi_1 + \phi_2}{2}$, $\phi_e = 2\pi\frac{\Phi}{\Phi_0}$ is the normalized external flux. α_1 and α_2 are different anharmonicity parameters of CPR in different JJ. In the case of $\phi_e = 0$, one can easily come to the conditions responsible for the double-well energy potential formation. If both JJs (with the same CPR) are of the same size the energy potential remains always symmetric. For this reason, any state-dependent current is impossible even if an external magnetic field is applied. However, at different sizes of JJ (with different critical currents) the external magnetic field always breaks the potential symmetry and produces a state-dependent current in the loop. In papers Amin et al. (2005), Klenov et al. (2010) influence of external magnetic field on the splitting of energy levels are investigated. It was shown that the ratio E_J/E_C also influence on splitting of energy levels. It is found that the value of

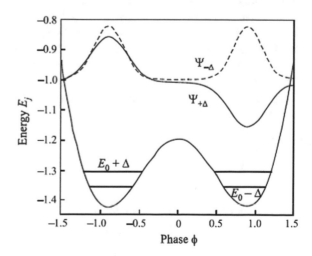

Fig. 4.17 Two-hump potential for silent qubit on JJ with anharmonic CPR

Fig. 4.18 Scanning-electron
micrograph of silent qubit on
JJ with anharmonic CPR
(Amin et al. 2005)

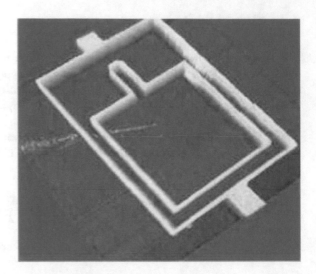

Φ_e, in which splitting parameter remains unchangeable. Silent qubit based on JJ with anharmonic CPR firstly was realized in study Amin et al. (2005). Scanning-electron micrograph of silent qubit presented in Fig. 4.18.

4.4   Flux Qubit with Three JJ

One of the disadvantage of a flux qubit with a single JJ concerns the large inductance of superconducting ring, the energy of which must be at the level of the Josephson energy to form the required double-well potential profile. It means that a large size of superconducting ring, which makes sensitive to dephasing by magnetic fluctuations of the environment (Wendin and Shumeiko 2007; Omelyanchouk et al. 2013). One way of solving this problem was suggested by Orlando et al. (1999). It was shown that replacing a large ring inductance by the Josephson inductance of an additional JJ, as shown in Fig. 4.20. In this figure, three JJs connected in series in the qubit ring. The inductive energy of the loop is chosen to be much smaller than the Josephson energy of JJs. The two junctions are identical while the third junction has smaller area, and therefore smaller critical current. Corresponding Hamiltonian has the form

$$H = E_C \left\{ n_1^2 + n_2^2 + \frac{n_3^2}{\alpha + 0.5} \right\} - E_J \left\{ 2 + \alpha - \cos(\phi_1) - \cos(\phi_2) - \alpha \cos(\phi_3) \right\},$$

$$(4.22)$$

where α refers to the ratio of critical current in three junction qubit (Fig. 4.19).

For the calculation of the energy the inductance of the loop is considered negligible, $\ell \ll 1$, so that the total flux is equal to the external flux. In this case, flux

Fig. 4.19 Schematic
representation of flux qubit
with three JJ

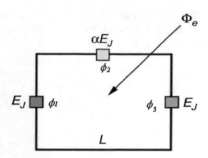

quantization around the loop containing the junctions gives $2\pi(f + 1/2)$. Here f is
the magnetic flux frustration and is the amount of external magnetic flux in the loop
in units of the flux quantum Φ_0

$$\phi_3 - \frac{2\pi\Phi}{\Phi_0} \approx \frac{2\pi\Phi_e}{\Phi_0} = 2\pi(f + 1/2). \tag{4.23}$$

If we take $\phi_e = \pi$ by introducing new variables, $\phi_\pm = (\phi_1 \pm \phi_2)/2$, the two-
dimensional periodic potential landscape of this circuit contains the double well
structures near the points $(\phi_+, \phi_-) = (0, 0)$ mod 2π (Fig. 4.20). An approximate
form of the potential energy structures is given by

$$U(\phi_+, 0) \approx E_J(-2\alpha\phi_+^2 + \phi_+^4/4). \tag{4.24}$$

Each well in this structure corresponds to clockwise and counterclockwise currents circulating in the loop. The amplitude of the structure is given by the parameter
αE_J, and for $\alpha \gg 1$ the tunneling between these wells dominates. Thus this qubit is
qualitatively similar to the single-JJ qubit described above, but the quantitative parameters are different and can be significantly optimized. For this qubit, single-qubit

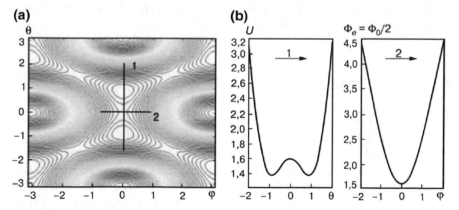

Fig. 4.20 Two-dimensional periodic potential landscape of three JJ flux qubit (**a**); Changing profile
of potential in direction (*1*) and (*2*) (**b**)

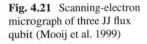

Fig. 4.21 Scanning-electron micrograph of three JJ flux qubit (Mooij et al. 1999)

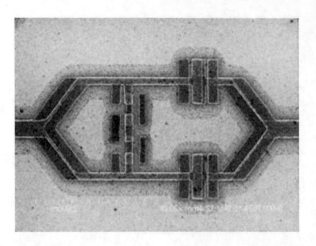

rotation (Rabi oscillation), the direct coupling between two qubits, and the entangled states have been demonstrated in Majer et al. (2005), Izmalkov et al. (2004), Chiorescu et al. (2004). In papers Ioffe et al. (1999), Blatter et al. (2001), Yamashita et al. (2006b) was proposed a qubit consisting of a superconducting loop with two zero JJs and one π JJ. In the system degenerate states appear in the phase space with $\phi_e = 0$ because of a competition between the zero and π states. Quantum tunneling between the degenerate states leads to a formation of bonding and antibonding states which are used as a bit. For manipulating the states only a small magnetic field around zero is required. This peculiarity leads to a large-scale integration and a construction of the smaller-sized robust qubit. Rabi oscillations in π junction embedded three JJ system was experimentally realized in Feofanov et al. (2010), Izmalkov et al. (2008). It was demonstrated a significant reduction in a size of the digital circuits (See also Likharev 2012). In contrast to π junctions based on HTS JJs, using SFS elements leads to large critical current and seems more compact. In this technology, it is not required to trap integer number of flux quanta in their superconducting loops. Flux qubit on three JJ firstly was realized in study Mooij et al. (1999). Scanning-electron micrograph of three junction qubit presented in Fig. 4.21.

4.5   Adiabatic Measurements of Superconducting Qubits

In this section the adiabatic measurements of three JJ qubits (Orlando et al. 1999) is considered. A flux qubit weakly coupled to a classical oscillator consisting of an inductor L_T and a capacitor C_T forming a tank circuit proposed in Greenberg et al. (2002) (Fig. 4.22). It was proposed to investigate flux qubits by the measurement of impedance, which is widely used to determine the CPR in JJs. It was analyzed in detail the case of a high-quality tank circuit coupled to a persistent-current qubit, to which the impedance measurement technique was successfully applied in the clas-

Fig. 4.22 Three JJ flux qubit adiabatically coupled with tank circuits

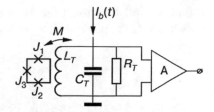

sical regime. It is shown that the low-frequency impedance measurement technique can give considerable information about the level anticrossing, in particular the value of the tunneling amplitude. As shown in Sect. 4.4, the circuit of flux qubit on three JJ is characterized (Orlando et al. 1999) by a two dimensional potential $U(\phi_1, \phi_2)$ which, for suitable qubit parameters, exhibits two minima. In the classical case these minima correspond to persistent current in the loop. If the applied magnetic flux $\Phi_x = \Phi_0/2$ both minima have the same potential, leading to a degenerate ground state. Consequently such qubit can be described by the Hamiltonian (Izmalkov et al. 2004)

$$H(t) \approx -\Delta\sigma_x - \varepsilon\sigma_z + A\cos(\omega t)\sigma_z, \qquad (4.25)$$

where σ_x, σ_z are the Pauli matrices for the spin basis and Δ is the tunneling amplitude. The qubit bias is given by $\varepsilon = I_p(\Phi_x - \Phi_0/2)$, where I_p is the magnitude of the qubit persistent current. The last term describes the microwave irradiation necessary for the spectroscopy. Due to the mutual inductance M the tank biases the qubit resulting $\Phi_{in} = \Phi_{dc} + \Phi_{rf}$.

As studied in Izmalkov et al. (2004), Greenberg et al. (2002) the renormalized resonant frequency of the tank-qubit system is given as

$$\omega_0 = \sqrt{\frac{k^*}{m^*}} \approx \omega_T \left\{ 1 - \frac{M^2}{2L_T} \frac{d^2 E_-(\Phi_{dc})}{d\Phi_{dc}^2} \right\}, \qquad (4.26)$$

which contains information on the curvature of the ground state of the qubit. The eigenvalues of the Hamiltonian (4.26) depend on the flux $\Phi_{bias} = \Phi_x - \Phi_0/2$:

$$E_\pm(\Phi) = \pm\sqrt{(I_p\Phi)^2 + \Delta^2}. \qquad (4.27)$$

Last equation showed that Δ and I_p can be determined from the dependence of the resonance frequency of the tank circuit as function of applied flux. Experimentally, the shift of the resonance frequency can be obtained by driving the tank circuit with a AC current I_{rf} at a frequency close to the resonant frequency ω_T and measure the phase shift θ between the AC voltage and driving current. For a small qubit inductance L, the phase shift is defined in study Greenberg et al. (2002)

$$\tan\theta = \frac{M^2 Q}{L_T} \frac{d^2 E_-(\Phi_{dc})}{d\Phi_{dc}^2}. \qquad (4.28)$$

The dependence of phase shift between tank circuit voltage and bias current on the flux bias is illustrated in Fig. 4.23. The solid lines are experimental data fitted by the theoretical curves (dashed curves) for qubit parameters $I_p = 225\,\text{nA}$ and $\Delta/h = 1.75\,\text{GHz}$. The curves correspond to various values of the AC-bias current on the tank circuit resulting in AC voltage amplitudes, from top to bottom, $V_T = 4.3, 2.9, 0.5,$ and $0.3\,\mu\text{V}$. Energy gap E between the qubit energy levels determined from the positions of the mid points of the peak and dip structures (solid squares). The solid line is the theoretical curve calculated from Eq. (4.28) using the parameters $I_p = 225\,\text{nA}$ and $\Delta/h = 1.75\,\text{GHz}$ obtained from the ground-state measurements. The effective temperature $T \approx 70\,\text{mK} \approx 1.4\,\text{GHz} \cdot h/k_B$ is smaller than the minimal energy 2Δ (Fig. 4.24).

Fig. 4.23 Dependence of phase shift between tank voltage and flux bias

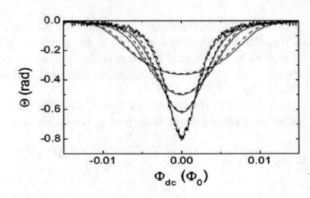

Fig. 4.24 Energy gap between the qubit energy levels as function of external magnetic flux

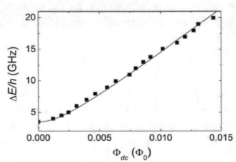

4.6 Direct Detection of Rabi Oscillation in Superconducting Qubits

The dynamics of two-level quantum system described using Bloch sphere presentation (Omelyanchouk et al. 2013). Any points at the end vector on the Bloch sphere corresponds to superposition of two quantum states $|0\rangle$ and $|1\rangle$ (Fig. 4.25). Rotation of vector at the Bloch sphere corresponds to evolution of qubit state. Dynamics of two-level qubit described by free precession of Bloch vector around axis z with frequency $\omega_{10} = \frac{E_1 - E_0}{\hbar}$ (Fig. 4.25a). The influence of dissipation effects in two-level systems leads to the trajectory on the Bloch sphere presented in Fig. 4.25b.

Rabi oscillations are coherent periodic transitions between the states of a two-level quantum system with a low Rabi frequency Ω_R,

$$\Omega_R = \sqrt{A^2 + (\Delta\omega)^2}, \tag{4.29}$$

induced by a harmonic field in resonance with the much larger inter level spacing $\Omega \gg \Omega_R$ (Zubayri and Sculli 2003). In last Eq. (4.29), $\Delta\omega$ means detuning of external frequency, A is the microwave power applied to qubit. They are among the most direct signatures of quantum behavior. Their observation in superconducting qubits was therefore a critical step in the direct proof of the qubits behavior and in the evaluation of control parameters. The decay rate of Rabi oscillations is determined by the relaxation and dephasing rates of the system and is quite fast, which makes their observation a nontrivial task (Omelyanchouk et al. 2013).

Rabi oscillation of qubit consisting of three JJs was carried out in study Chiorescu et al. (2004). Rabi oscillations for a resonant frequency $f = \omega_{10} = 6.6\,\text{GHz}$ and three different microwave powers $A = 0, -6$ and $-12\,\text{dB}$, where A is the nominal

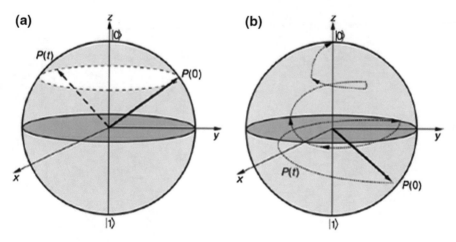

Fig. 4.25 Dynamic of two-level qubit system on the Bloch sphere: the case of absence of dissipation (**a**); the case of inclusion of dissipation (**b**)

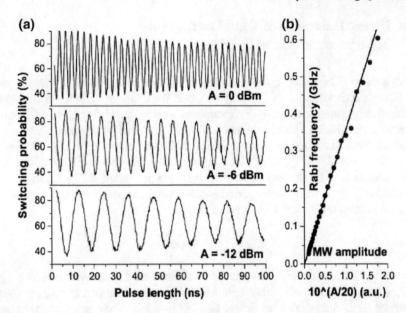

Fig. 4.26 Rabi oscillation of three JJ qubit (**a**). Linear dependence of the Rabi frequency on the microwave amplitude (**b**)

microwave power applied at room temperature (Fig. 4.26a). The data are well fitted by exponentially damped sinusoidal oscillations. The resulting decay time is ∼150 ns for all powers. Linear dependence of the Rabi frequency on the microwave amplitude, expressed as $10^{A/20}$. The slope is in good agreement with estimations based on circuit design, fabricated in Chiorescu et al. (2004).

Chapter 5
Chaotic Phenomena in Josephson Systems

Abstract In this chapter, we present some results of chaotic behavior of JJ systems. Firstly we discuss physical foundation and characteristics of chaotic dynamics of single JJ with generalized CPR, including anharmonic effects. Influence of control parameters on the dynamics of fractal JJ are discussed in second section. In the next section of the chapter, chaotic dynamics of resistively coupled two JJ system is presented. We further explore the system parameters and focus on the regions where a higher complexity is encountered. In Sect. 5.4 it will be shown that a simpler DC-driven JJ device can generate a wide region of hyperchaos without using an AC source compared to other studies existing in the literature. By doing so, the sensitivity of superconducting devices without the negative impacts of RF-excitation is combined with the highly-chaotic regimes in order to model a device with higher unpredictability. Last section is related with the discussion of synchronization of chaos in system of JJ with resistive coupling.

5.1 Chaotic Dynamics of a Single JJ and the Influence of Anharmonic CPR

It is well known that many simple nonlinear systems, including JJs, can show chaotic dynamics. While the equations of motion of such systems are deterministic, the solutions, when chaotic, include a pseudo-random component which is, for practical purposes, indistinguishable from noise. In Josephson systems chaotic noise can exceed thermal noise by several orders of magnitude, making the possibility of chaos an important consideration in the design of low-noise circuits. From this point JJ devices could be useful for ultrahigh-speed chaotic generators for applications of code generation in spread-spectrum communications (Kennedy et al. 2000) and true random number generation in secure communication and encryption (Sugiura et al. 2011; Uchida et al. 2003). From this point, the dynamics of JJs is of great importance in superconducting electronics (Dana et al. 2001) so that junction devices can be used as chaotic generators.

In any chaotic secure communication system, a chaotic signal is used to mask the message to be transmitted. As known (Sprott 2003; Parker and Chua 1987), the

© Springer International Publishing AG 2017
I. Askerzade et al., *Modern Aspects of Josephson Dynamics and Superconductivity Electronics*, Mathematical Engineering,
DOI 10.1007/978-3-319-48433-4_5

ordinary chaotic systems have one positive Lyapunov exponent and this exponent helps to mask the message. However, as shown in Perez and Cerdeira (1995) the messages masked by such a normal chaotic system were not always safe. In study Pecora (1996) it was found that this problem could be overcome by using a higher-dimensional hyperchaotic system, which has an increasing randomness and higher unpredictability. In addition, the application and practical aspects of a chaotic random number generator (Stojanovski et al. 2001) can further be improved by hyperchaos. Hyperchaotic dynamics has attracted at great interest in the last few years, in the fields of nonlinear circuits (Cafagna and Grassi 2003; Grassi and Mascolo 2002), secure communications (Udaltsov et al. 2003; Hsieh et al. 1999) and synchronization.

In literature, different models of JJ is widely used to understand whether such a superconducting junction can be applied as a transmitter and receiver in chaotic digital communications. In this context, different models of JJ, namely, shunted linear resistive-capacitive junction (RCSJ), shunted nonlinear resistive-capacitive junction and the shunted nonlinear resistive-capacitive-inductance junction (RCLSJ) have briefly been reviewed in this section. The first two models show chaotic behaviors when driven by external sinusoidal signal while the shunted inductance junction generates chaotic oscillation with external DC bias only (Kautz 1981). The RCLSJ model is found more appropriate (Huberman et al. 1980; Kornev and Semenov 1983; Edward 1993) in high-frequency applications (see Fig. 1.19 and Eqs. (1.59)–(1.62)). In the framework of RCLSJ junction model, the chaotic oscillation has been modulated in response to both the amplitude and frequency of an external sinusoidal signal. This external signal may be assumed as information signal for transmission in chaotic system.

The resistively, capacitively, inductively shunted JJ (RCLSJ) circuit model shown in Fig. 1.19 ($R_s \ll R_N$). In Eq. (1.60) R_N and R_s refer to normal state and shunt resistances respectively. The system of equations describing the single JJ in Fig. 1.19 is given by Eq. (1.62), where I_s denotes the current in the shunt branch and the voltage dependent junction resistance given by Eq. (1.59). $V_g = 2\Delta/e$ in Eq. (1.59) is the gap voltage that depends on the energy gap (i.e., Δ) of JJ, R_N is the normal state resistance and R_{sg} is the sub-gap resistance of the JJ in the superconducting state. In principle, the white noise can be included into the right hand side of Eq. (1.60) in the form of delta function. The strength of normalized fluctuations is proportional to the ratio of thermal energy to Josephson energy: $\gamma_T = 2\pi k_B T/I_c \Phi_0$. Such ratio becomes negligibly small (i.e., $\gamma_T \ll 1$) for JJs with high critical current. Detailed description of influence of thermal fluctuation on JJ dynamics presented in books Likharev (1986); Barone and Paterno (1982). That is the reason why we ignore the effect of thermal noise in our model.

The system of third order autonomous equations (1.59)–(1.62) can be reduced to a dimensionless form so that we perform chaotic dynamical analysis for various control parameters (Canturk and Askerzade 2013)

$$\dot{\phi} = v \tag{5.1}$$

$$\dot{v} = \frac{1}{\beta_C} (i_{dc} - g(v)v - f(\phi) - i_s) \tag{5.2}$$

$$\dot{i_s} = \frac{1}{\beta_L} (v - i_s) \tag{5.3}$$

where $\beta_C = (2e/\hbar)I_c C R_s^2$ is the McCumber parameter (McCumber 1968); $g(v) = R_s/R(v)$ is the normalized tunnel junction conductance; $i_s = I_s/I_c$ is dimensionless shunt current; $i_{dc} = I_{dc}/I_c$ is dimensionless external DC bias current; $\beta_L = (2e/\hbar)I_c L_s$ is the dimensionless inductance; $\tau = \omega_c t$ is the normalized time in which $\omega_c = (2e/\hbar)V_c$ is the characteristic frequency and $V_c = I_c R_s$ is the characteristic voltage; $f(\phi) = (\sin\phi - \alpha\sin 2\phi)$ with anharmonicity parameter α (See Sect. 1.5 for more detail). The relationship between β_C and ω_c can be given by $\beta_C = (\omega_c/\Omega_{p0})^2$ with the help of plasma frequency $\Omega_{p0} = \sqrt{2eI_c/\hbar C}$.

In practical circuit design, it is useful to have a detailed knowledge about the dynamical behavior of the system in different parameter ranges, so that one can avoid certain undesirable regions in the parameter space to optimize the circuit performance. For that purpose, bifurcation diagrams are very useful tools for visualizing the global dynamical behavior of a system. In this section, bifurcation diagrams will be determined under the influence of the anharmonic CPR to illustrate various oscillations that are present in our system. The quantitative criterion for a time series to be chaotic is the positivity of one of its Lyapunov exponents (Sprott 2003; Parker and Chua 1987). In this section, we have three independent variables in RCLSJ model and we will compute only the largest Lyapunov exponent, which is sufficient to distinguish a chaotic signal from a periodic signal. Figure 5.1 shows a bifurcation diagrams obtained using local maxima method, while changing the control parameter β_L in the range $0 \leq \beta_L \leq 10$ for different α values (Canturk and Askerzade 2013). The other parameters are held constant at values $\beta_C = 0.707$ and $i_{dc} = 1.2$ similar to Whan et al. (1995). As shown in the Fig. 5.1b, the presence of second harmonics in CPR leads to a change in the bifurcation diagrams: compared to Fig. 5.1a, the first beginning of the chaotic region (i.e., between $\beta_L \approx 2.2$ and $\beta_L \approx 3$ at $\alpha = 0$) in Fig. 5.1a has shifted to (i.e., between $\beta_L \approx 1.5$ and $\beta_L \approx 1.8$ at $\alpha = 0.5$). Besides, the width of the chaotic region becomes smaller. Such behavior also observed for different α values. Furthermore, the distance between the first chaotic region and the second one at $\alpha = 0$ decreases for non-zero α values. Corresponding largest Lyapunov exponent of the RCLSJ circuit with and without anharmonic CPR are presented in Fig. 5.2.

In order to interpret the preceding results for Fig. 5.1, we use the same terminology discussed in Canturk and Askerzade (2011) for JJ with anharmonicity CPR. From this study, second harmonics in CPR leads to an increase in the normalized critical current: I_c/I_{c_0} where I_{c_0} is the critical current at $\alpha = 0$. Therefore, it causes to a decrease in the inductance of the JJ

$$L_J(\alpha) = \frac{\hbar}{2e} \frac{1}{I_c(\alpha)} \tag{5.4}$$

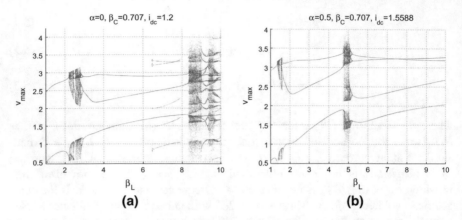

Fig. 5.1 Bifurcation diagram of JJ with harmonic (**a**) and anharmonic (**b**) CPR

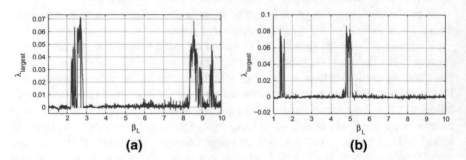

Fig. 5.2 Lyapunov exponents for JJ with harmonic (**a**) and anharmonic (**b**) CPR

It is well known that the influence of shunted inductance L_s (i.e. $\beta_L = (2e/\hbar)I_cL_s$) on the dynamics of JJ is important if it is comparable with Josephson inductance (i.e., $L_s \approx L_J$) (see Fig. 5.2). Consequently, the values of the shunted inductance L_s together with L_J has shifted to lower values (see Fig. 5.1) and it leads to the above mentioned changes in bifurcation diagrams of the JJ with anharmonic CPR.

Determining an appropriate chaotic regions from bifurcation diagrams in Fig. 5.1 as well as associated Lyapunov exponents in Fig. 5.2, we can obtain time series representations in Fig. 5.3. Corresponding phase portrait is illustrated in Fig. 5.4.

5.2 Chaotic Behavior of Fractal JJ

Insulating layer of an ideal tunnel JJ is considered to be homogeneous. However, the formation of the layer in the lithographic process is inevitable to have intrinsic fractal substructures (Kruchinin et al. 2005, 2008). In the lithographic superconducting device, it is possible to make the circuits that have a variety of different design

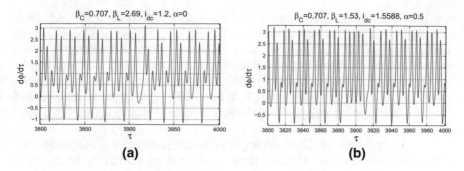

Fig. 5.3 Time series for JJ with harmonic (**a**) and anharmonics (**b**) CPR

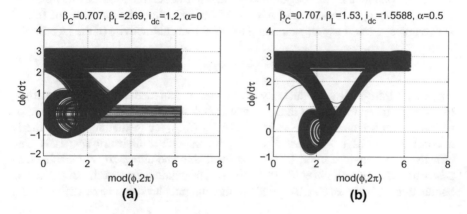

Fig. 5.4 Phase portrait of JJ with harmonic (**a**) and anharmonic (**b**) CPR

parameters compared with semi-conducting circuits, the construction of superconducting circuits has a considerable manufacturing variability. Each lithographic JJ
in an integrated circuit will have slightly different properties such properties are
assumed to have a fractal nature. Depending on fractal dimensions and porosity of
materials, the macroscopic properties of them are characterized by heat and elec-
tric conductivities, diffusion and etc. (Kitamura and Yokoyama 1991). Recently,
nanoporous organic silicon composite materials were prepared by taking advantage of low dielectric constant as porous materials (Xi-Jie et al. 2010; Yu and Li
2001). The nonlinearity and memory properties of fractal junctions allow us to use
such a contact as an effective, easy to modify generator of chaotic signals with
a controllable statistical structure. Such a generator can serve as the technological basis for the chaotic communication (Kruchinin et al. 2005). As mentioned
before, JJ devices could be useful for ultrahigh–speed chaotic signal generators
for applications of code generation in spread–spectrum communications (Kennedy
et al. 2000) and true random number generation in secure communication and
encryption (Sugiura et al. 2011; Uchida et al. 2003). As noted in Sect. 5.1, chaos
in conventional JJs and Josephson circuits was studied by many authors, see for

example (Kautz 1981; Huberman et al. 1980; Kornev and Semenov 1983). Although, some aspects of the chaotic dynamics of the fractal junctions was conducted by Kruchinin et al. (2005, 2008) for restrictive values of control parameters, we believe needs the detailed analysis of junction dynamics and chaotic behavior. This implies that the detailed behavior of chaotic dynamics in terms of control parameters remains interesting. Contrary to studies (Kruchinin et al. 2005, 2008), presented investigation for a fractal JJ is based on the analysis of Lyapunov exponents and bifurcation diagrams.

The literature review exploits the importance of the influence of the fractal characteristics of non-conductive layer on the physical properties of the JJs. In the present section, we conduct the influence of the fractal parameters of the chaotic dynamics for various parameters. Investigation presented in this section is based on numerical simulations using nonlinear conductivity of fractal JJ. For the modeling of fractal JJ dynamics, we will use an expression for effective conductivity of the layer in the form

$$\sigma = \sigma_e + \alpha_e E^2, \tag{5.5}$$

where σ_e is the effective conductivity of two component medium and α_e is the effective nonlinear conductivity of the layer which is described by the physical properties of fractal substructures. In the derivation of Eq. (5.5), the insulating layer of the JJ is considered to have a mixture of conducting as well as dielectric materials. The effective conductivities in Eq. (5.5) are controllable and allows to get tunable characteristics of JJ (Kruchinin et al. 2005, 2008). The detailed analysis for 2D layer, the conductivity of such an insulator at phenomenological level can be described by

$$\sigma \simeq \sigma_m \xi^{\frac{1}{2}} + \frac{\alpha_D}{\xi^{\frac{3}{2}}} E^2, \tag{5.6}$$

where $\xi = \sigma_D / \sigma_m$ is the ratio of conductivity of dielectric component σ_D to the conductivity of metal σ_m the mixture and α_D is the nonlinear conductivity of dielectric part in the mixture Kruchinin et al. (2005, 2008).

The influence of fractal effects on the JJ dynamics is equivalent to introducing effective junction resistance. Therefore, the inverse of the fractal resistance $1/R$ can be written in terms of σ/σ_m

$$\frac{1}{R} = \frac{1}{R_0} \frac{\sigma}{\sigma_m} = \frac{1}{R_0} \left[\xi^{\frac{1}{2}} + \frac{\alpha_0}{\xi^{\frac{3}{2}}} \dot{\phi}^2 \right], \tag{5.7}$$

where R_0 is assumed to be the resistance of an ideal junction; $\alpha_0 = \left(\frac{\Phi_0}{2\pi d} \right)^2 \alpha_D \sigma_m$ and d is the thickness of the junction. After substituting Eq. (5.7) into Eq. (5.3), one can obtain the dimensionless form of the equation

$$\beta_C \ddot{\phi} + \left[\xi^{\frac{1}{2}} + \frac{\alpha_0}{\xi^{\frac{3}{2}}} \dot{\phi}^2 \right] \dot{\phi} + \sin \phi = i_\omega \sin(\omega t). \tag{5.8}$$

Notations in Eq. (5.8) are the same as in Eqs. (5.1)–(5.3). The solutions of Eq. (5.8) are numerically obtained by using adaptive Runge–Kutta method.

As noted above, in practical circuit design, it is useful to have a detailed knowledge about the dynamical behavior of the system from different perspectives so that one can avoid certain undesirable regions in the parameter space to optimize the circuit performance. We analyze of the fractal junction with the plot of bifurcation diagram and an associated largest Lyapunov exponent in Fig. 5.5. The bifurcation diagram in the figure is obtained by using local maxima method for voltage $v = \dot{\phi}$, while changing the control parameter $\omega = \Omega/\omega_c$ (Ω refers to frequency of driving force) in the range $0 \leq \omega \leq 0.5$ for a fractal JJ at constant $\beta_C = 25$, $\xi = 0.5$, $\alpha = 0.02$ and $i_\omega = 1.5$ values (Canturk and Askerzade 2013; Kruchinin et al. 2005). In the present section, the DC bias current I_{dc} was taken to be zero for all computations. The influence of non-zero DC bias together with AC source is open subject.

Bifurcation diagram is useful tools for assessing the characteristics of the steady-state solutions of a system over the range of control parameter. As shown in Fig. 5.5, the dynamics is complex for $\omega < 0.25$ while the dotted single lines for $0.25 \leq \omega$ implies a pure AC Josephson oscillations. As shown in Fig. 5.5, the plot of the largest Lyapunov exponents which is also useful to visualize the global dynamical behavior, is complementary to the plot in bifurcation diagram. Lyapunov exponents are obtained in order to check whether the time series to be chaotic by looking at the positivity of one of its Lyapunov exponents (Sprott 2003). The algorithm used to determine the Lyapunov exponents is based on Wolf et al. (1985b). In the plot, $\lambda_{\text{largest}} \leq 0$ corresponds to the periodic motion in the junction. The findings in these figure look similar to the one studied in Canturk and Askerzade (2013), Wolf et al. (1985b).

In order to describe the detailed knowledge about the dynamical behavior of the system in the plane of normalized amplitude of an external AC source $i_\omega(\tau) = I_\omega/I_c$ versus normalized frequency $\omega = \Omega/\omega_c$, the contour plots are illustrated in Figs. 5.6 and 5.7. Figure 5.6 corresponds to an ideal JJ while Fig. 5.7 corresponds to a fractal junction. For each figure there exist three contour plots for different McCumber

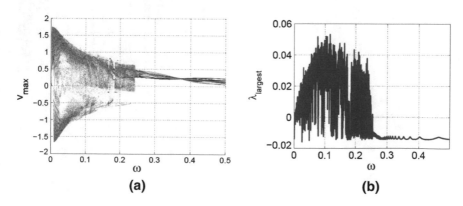

Fig. 5.5 Bifurcation diagram (*left*) and largest Lyapunov exponent (*right*) of fractal JJ

Fig. 5.6 Mapping of chaotic
and oscillation regions of
normal (nonfractal) JJ with
harmonic CPR

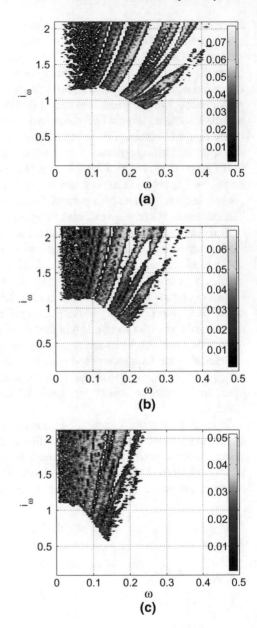

Fig. 5.7 Mapping of chaotic
and oscillation regions of
fractal JJ

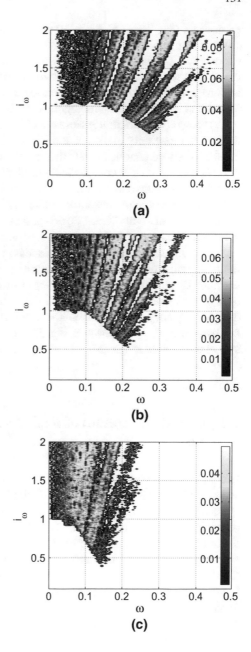

(a)

(b)

(c)

parameters (i.e., $\beta_C = 5$, $\beta_C = 10$ and $\beta_C = 25$). Each contour plot has a vertical color bar on the right to indicate the magnitude of the largest positive Lyapunov exponent λ_{largest}. In other words, $\lambda_{\text{largest}} > 0$ with different magnitude corresponds to the chaotic regions represented in colored islands on 2D plane. However, the white regions in these plots corresponds periodic motions that are represented as a single dotted line in the bifurcation diagrams shown in Fig. 5.5 for $i_\omega = 1.5$.

As shown (Canturk and Askerzade 2014) in Figs. 5.6 and 5.7, the increase in β_C leads to homogeneity in the chaotic islands. However, the decrease in β_C leads to a number of sliced chaotic islands that are separated by pure oscillation regions. The influence of β_C on the contour plots implies the contribution of this parameter on the second derivative of Eq. (5.8). When we compare Figs. 5.6 and 5.7, the chaotic islands and periodic regions look similar in shape but differs in size. In other words, the contour plots with fractal layer seems to be the expanded form of the contour plots obtained for an ideal junction. The contribution of fractal effect in the junction on the homogeneity of the chaotic islands on the contour plots can be explained by the renormalization of the junction resistance given in Eq. (5.6). When we analyze the contour plots vertically, the increase in β_C leads to a lowering of the amplitude of the external drive in plane (i_ω, ω). In the horizontal direction the effect of driving frequency leads to the contour map suppression with increasing of McCumber parameter β_C. As shown in Fig. 5.6, the influence of the fractal effects on the size of bifurcation range seems to be equivalent to the reduced effective McCumber parameter.

5.3 Chaotic Dynamics of Resistively Coupled DC-Driven JJs

In contrary to previous sections, in this section two distinct JJs connected with a constant coupling resistance R_{cp} are theoretically considered to investigate the overall dynamics below and above the critical current I_c. The circuit model of the device is driven by two DC current sources, I_1 and I_2. Each junction is characterized by a nonlinear resistive capacitive junction (NRCSJ). Having constructed the circuit model, time-dependent simulations are carried out for a variety of control parameter sets. Common techniques such as bifurcation diagrams, two-dimensional attractors and Lyapunov exponents are applied for the determination of chaotic as well as periodic dynamics of JJ devices. The chaotic current which flows through R_{cp} exhibits a very rich behavior depending on the source currents I_1 and I_2 and junction capacitances C_1 and C_2. The device characteristics are summarized by a number of three-dimensional phase diagrams in the parameter space. In addition, for certain parameters, hyperchaotic cases with two positive Lyapunov exponents are encountered, which is the subject of following Sect. 5.4.

We construct a model of two resistively coupled JJs, in which two non-identical nonlinear resistive capacitive junction (NRCSJ) circuits are connected to form the

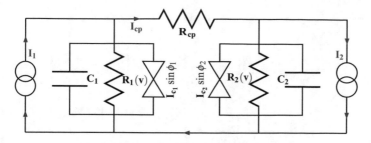

Fig. 5.8 Circuit model of resistively coupled DC-driven JJ

circuit shown in Fig. 5.8. The circuit equations for each NRCSJ can be obtained using Kirchoff's law (Kurt and Canturk 2009):

$$C_i \frac{dV_i}{dt} + \frac{V_i}{R_i(V_i)} + I_{c_i} \sin \phi_i = I_i - I_{cp} \tag{5.9}$$

$$\frac{\hbar}{2e} \frac{d\phi_i}{dt} = V_i \tag{5.10}$$

$$I_{cp} = (V_1 + V_2)/R_{cp} \tag{5.11}$$

Here, V_i is the voltage across the ith junction ($i = 1, 2$), I_i is DC current source, I_{c_i} is the critical junction current, and C_i represents the junction capacitance of the ith junction. In principle, phase derivatives are of central importance because they are proportional to the junction voltages (see Eq. (5.11)). Note that the current I_{cp} through the coupling resistance R_{cp} is given by Eq. (5.11) in terms of voltage V_i on each junction. It can be seen that the coupling arises as a natural consequence of the exchange of current through the resistor R_{cp} and it depends on the total voltage on the junctions.

Initially, we should state that the coupling resistance R_{cp} plays an important role for the desired regime of the system. In fact, the differential terms cannot be negligible for the specified parameter range. Now, we wish to justify this point and prove that the nonlinear resistance $r_1(v_1)$ cannot always drive the system alone, but R_{cp} also contributes to the dynamics of the system. System of Eqs. (5.9)–(5.11) reduced to following system of equations after the substitution of the coupling current expression. Note that the second term not only includes $r_1(v_1)$ but it also includes the parameter R_{cp} as $\gamma(v_1 + v_2)$. System of Eqs. (5.9) and (5.11), the non-dimensional system equations has a form

$$\dot{v}_1 + g_1(v_1)v_1 + \sin \phi_1 = \{i_1 - \gamma(v_1 + v_2)\}, \tag{5.12}$$

$$\dot{\phi}_1 = v_1, \tag{5.13}$$

$$\dot{v}_2 + g_2(v_2)v_2 + \Gamma \sin \phi_2 = \alpha \{i_2 - \gamma(v_1 + v_2)\}, \tag{5.14}$$

$$\dot{\phi}_2 = v_2, \tag{5.15}$$

where

$$i_1 = \frac{I_1}{I_{c_1}}, \quad i_2 = \frac{I_2}{I_{c_1}}, \quad \gamma = \frac{V_0}{I_{c_1} R_{cp}}, \alpha = \frac{C_1}{C_2}, \quad \Gamma = \frac{I_{c_2}}{I_{c_1}} \frac{C_1}{C_2}, \tag{5.16}$$

$$g_1(v_1) = \frac{V_0}{I_{c_1} R_1(v_1)} = \begin{cases} \frac{V_0}{I_{c_1} R_1^n} & \text{if } v_1 > V_g/V_0 \\[2mm] \frac{V_0}{I_{c_1} R_1^{sg}} & \text{if } v_1 \leq V_g/V_0 \end{cases},$$

and

$$g_2(v_2) = \frac{C_1}{C_2} \frac{V_0}{I_{c_1} R_2(v_2)} = \begin{cases} \frac{C_1}{C_2} \frac{V_0}{I_{c_1} R_2^n} & \text{if } v_2 > V_g/V_0 \\[2mm] \frac{C_1}{C_2} \frac{V_0}{I_{c_1} R_2^{sg}} & \text{if } v_2 \leq V_g/V_0 \end{cases}.$$

Besides,

$$R_1(v_1) = \begin{cases} 244 \;\; \Omega & \text{if } |v_1| < V_g/V_0 \\ 0.366 \; \Omega & \text{if } |v_1| > V_g/V_0 \end{cases}$$

$$R_2(v_2) = \begin{cases} 422 \;\; \Omega & \text{if } |v_2| < V_g/V_0 \\ 0.633 \; \Omega & \text{if } |v_2| > V_g/V_0 \end{cases}, \tag{5.17}$$

where $V_g = 2.26\,\text{mV}$ is considered. Nonlinear resistances of the circuit are obtained from an experimental study of Algul et al. (2007). The inverse of nonlinear resistance (i.e. $1/R_i$) is considered as a dimensionless damping factor of the dynamic system. In addition, two critical currents are selected from Algul et al. (2007) as $I_{c_1} = 125\,\mu\text{A}$ and $I_{c_2} = 190\,\mu\text{A}$ respectively. The quantity $1/R_i$ can also be considered as a dimensionless damping factor of the dynamic system. The numerical solutions of Eqs. (5.12)–(5.14) are obtained by the fourth-order Runge–Kutta method (Kurt and Canturk 2009). The initial conditions are specified as $(\phi_i(0), v_i(0)) = (0, 1)$. The coupling resistance R_{cp} connecting the two different NRCSJ circuits is chosen to be $R_{cp} = 1\,\text{k}\Omega$ for the entire study. Since we are dealing with the construction of attractors, bifurcation diagrams and Lyapunov exponent spectra, large data sets are obtained from the simulations. In most cases, 100 000 data points are evaluated for the estimation of the Lyapunov exponents by using different control parameter sets. In the case of bifurcation diagrams, sufficient numbers of data points are obtained from the simulations for increasing bifurcation parameters (Kurt and Canturk 2009).

Since we have two non-mutual junction devices, we deal with the resultant coupling current I_{cp} which flows through the coupling resistance R_{cp}. In this manner, Figs. 5.9a and 5.10a represent time series of phases for two junctions. In Fig. 5.9a, the time-dependent variation of the phase of junction 1 contains more complexity than that of junction 2, since the higher frequency and random fluctuations are clearly seen. However, this situation only depends on the parameters of the two junctions. If appropriate parameters are applied to one of the junctions, the complexity increases or shifts from junction 1 to junction 2. For instance, the complexity is higher than that

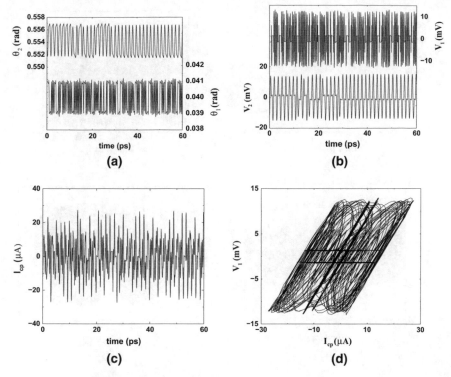

Fig. 5.9 **a** Phase differences (θ_1 and θ_2), **b** Voltages (V_1 and V_2), **c** Coupling current I_{cp} as a function of time and **d** Attractor for parameters $C_1 = 3.22fF$, $I_1 = 5\mu A$, $C_2 = 15fF$, $I_2 = 100\mu A$

of junction 1 in Fig. 5.10a. It can be stated that the effects of junction capacitances and the source currents play an important role in characterizing the phase dynamics. Strictly speaking, small source currents yield more chaotic oscillations for this parameter set without generalization. At this point, one should compare the fluctuations of junction 2 in Figs. 5.9a and 5.10a for two distinct current inputs. A detailed analysis has proven that the phases become larger by increasing the source current I_i. For instance, fluctuations can be observed around $\theta_1 = 0.04$ rad for $I_1 = 5$ mA and $\theta_1 = 412$ rad for $I_1 = 50$ mA. Note that the role of the junction capacitance can be negligible in this context. Figures 5.9b and 5.10b show the voltages V across the two junctions as a function of time. In principle, the first derivative of the phase gives the voltage in any JJ according to the second equation of Eqs. (5.13)–(5.15) when a single JJ is considered. Thus, whenever the fluctuation frequencies increase, the junction voltage has larger amplitudes. It can be stated that high-frequency oscillations in voltage V_1 of junction 1 give a useful medium for high-frequency signal generation. Note that one can easily adjust the frequency of the junction voltage by adjusting the source current I_i. For a better comparison, we refer to voltage V_1 in Fig. 5.10b since the frequency of the voltage signal in the figure is larger than in Fig. 5.9b. Furthermore, the frequency of V_2 also increases in the second case due

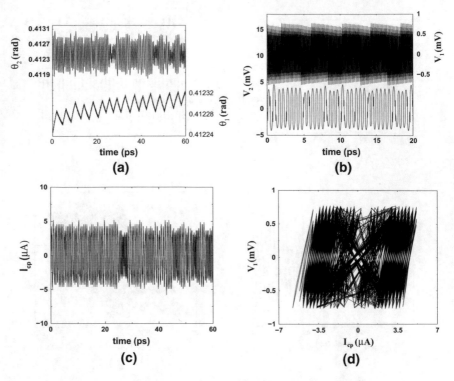

Fig. 5.10 a Phase differences (θ_1 and θ_2), **b** Voltages (V_1 and V_2), **c** Coupling current I_{cp} as a function of time and **d** Attractor for parameters $C_1 = 10fF$, $I_1 = 50\mu A$, $C_2 = 12fF$, $I_2 = 76.16\mu A$

to the increasing frequency of V_1, since the two junctions are coupled. The current I_{cp} through R_{cp} is shown in Figs. 5.9c and 5.10c for two representative parameter sets. Due to the connectivity of the junctions they contribute to the main current which flows through the coupling resistance. The chaotic characteristics is clearly seen for the two cases. The amplitude and frequency of the current can easily be adjusted by selecting the appropriate source currents in case of well-defined device parameters. For instance, the peak to peak amplitude of I_{cp} for the first parameter set is around 40 mA (in Fig. 5.9c), whereas it is around 10 mA (in Fig. 5.10c) for the second parameter set. The change in frequency is also observable from Figs. 5.9c to 5.10c. Figures 5.9d and 5.10d display the phase space constructions (i.e. attractors) of current I_{cp} versus voltage V_1. The effect of the nonlinear resistances is clearly seen in Fig. 5.9d. A tilted cross presumably indicates a sudden change in current when the definition of the nonlinear resistance in Eq. (5.18) is fulfilled. In addition, this effect cannot be seen clearly in Fig. 5.10d due to the high-frequency oscillations. In fact, phase space points produce four distinct regions in the I_{cp} V_1 plane in this figure.

In Figs. 5.11 and 5.12, some representative bifurcation diagrams are given as functions of capacitance and current. One can adopt a bifurcation analysis on both phase difference and voltage values as in Kurt and Canturk (2009). The bifurcation

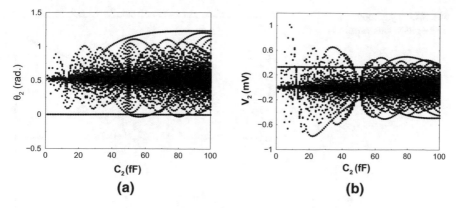

Fig. 5.11 Bifurcation diagrams of **a** the phase difference θ_2, **b** voltage V_2 as a function of junction capacitance C_2 for parameters: $I_1 = 62.5\mu A$, $I_2 = 95\mu A$ and $C_1 = 10fF$

diagram of ϕ_2 as a function of C_2 (Fig. 5.11a) indicates a rich dynamic behavior since the increasing capacitances yield more complex behavior in the system. Furthermore, the ϕ_2 values changes around 0.5 rad with C_2. There exist two periodic windows around $C_2 = 1$ and 13 fF. Similarly to the ϕ_2 values, there exist periodic behaviors at the same capacitances for voltage V_2 (Fig. 5.11b) for the same parameter set. The complexity in V_2 increases with an increase in capacitance, as in Fig. 5.11a. Figure 5.12 shows bifurcation diagrams as a function of the source current. In contrast to the previous diagram, the behavior of the system looks different in this case. Strictly speaking, the phase differences in Fig. 5.12a increase to higher values. The diagram represents a bouquet-like pattern. As a result of this trend, the volume of the pattern increases rapidly as a function I_2. In case of voltage V_2, the bifurcation diagram in Fig. 5.12b represents a more compact form in comparison to Fig. 5.11b. With increasing current, the shape of the diagram resembles a goblet-like pattern.

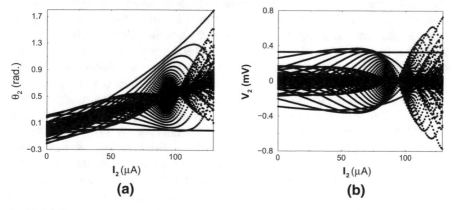

Fig. 5.12 Bifurcation diagrams of **a** the phase difference θ_2, **b** voltage V_2 as a function of source current I_2 for parameters: $I_1 = 62.5\mu A$, $C_1 = 10fF$, and $C_2 = 50fF$

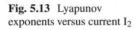

Fig. 5.13 Lyapunov
exponents versus current I_2

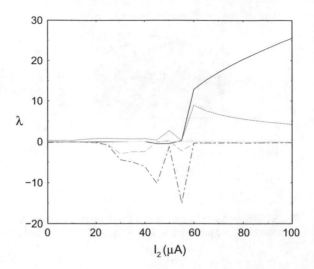

The volumes of the trajectories are restricted to a small value between 0.4 mV below $I_2 = 100$ mA; however, thereafter the volumes increase sharply (Kurt and Canturk 2009).

The rest of this section presents some numerical results on the Lyapunov spectra. As a reminder, it is very well known from the literature that the existence of at least one positive Lyapunov exponent is an indicator of chaotic behavior in the system (Wolf et al. 1985b). The dimension of the Lyapunov exponent is inferred from a number of dependent system variables (for instance, in our case Eqs. (5.9)–(5.11) implies four dimensions). The existence of a negative maximal Lyapunov exponent is an indicator of stable and fixed-point dynamic behavior. Similarly if a maximal exponent with λ_0 exists, then a periodic character is observed. Figure 5.13 indicates the Lyapunov spectra including four independent exponents with respect to current I_2. As indicated in the figure, the exponents are observed to be positive for all I_2. For small I_2 values, a slight increase in the maximal exponent is obvious. However, there exists a critical I_2 value (i.e. $I_2 = 55$ mA) at which the maximal exponent increases sharply. Note also that two exponents together have positive values for higher currents. This behavior shows a hyper-chaotic state just above $I_2 = 55$ mA; however, the other two exponents are found to be negative for higher currents.

Figure 5.14 shows some three-dimensional representations (i.e. 3D phase diagrams) of the maximal exponents as functions of the device parameters. Initially, a strong chaotic region is observed in Fig. 5.14a as a function of the junction capacitances. While a decrease in C_1 causes relatively higher exponents and chaoticity, the values of C_2 do not cause so much change for this parameter range. We believe that the change in with respect to C_1 is an exponential decrease. In addition, it is approximately constant with C_2. Since I_2 is very small compared to I_1, the capacitance of the second circuit could not play a major role dynamically. In order to verify the effect of the source currents (i.e. I_1 and I_2), we have repeated the simulation (shown

Fig. 5.14 3D phase diagram
of maximal exponents as
function of device
parameters

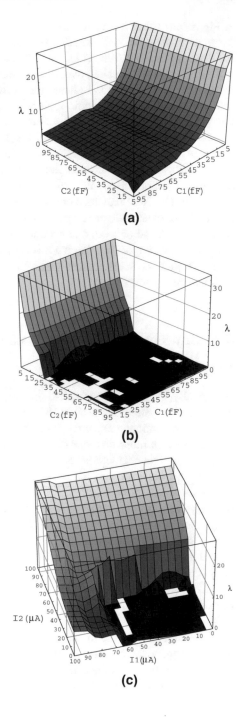

in Fig. 5.14a) for reversed source currents (i.e. $I_1 = 10\,\text{mA}$ and $I_2 = 80\,\text{mA}$). As in the previous case, it is seen that the change in with respect to C_2 is an exponential decrease in Fig. 5.14b. Thus, the effect of source current cannot be negligible in order to generate the desired dynamics of such a junction device. Note that the white regions in the C_1C_2 plane indicate the periodic regions for the applied parameter set. It is clear that the periodic regions mostly survive for small C_1 and moderate C_2 values. In addition, there exist some stable regions for high C_1, as indicated in Fig. 5.14b. The remaining 3D plot in Fig. 5.14c presents a different scenario. The effects of source currents are mainly indicated in this figure. Strictly speaking, high source currents yield more complex dynamics in the sense that the stable regions are found for relatively small current values with two independent islands (i.e. white stable regions). Note that each of two islands corresponds to small I_1 and I_2, respectively. There exist sudden jumps to higher values for moderate current values. In addition to this, some local exponential hills are encountered as a result of changes in the maximal exponents. Such a hill is also shown in Fig. 5.13. As a result, small values of junction capacitances yield larger Lyapunov exponents and complexity with increasing source currents.

5.4   Bifurcation and Hyperchaos of Resistively Coupled DC-Driven JJs

Historically, the need for hyperchaos comes from the field of secure communication. Hyperchaotic systems are usually classified as chaotic systems which have more than one positive Lyapunov exponent, indicating that the chaotic dynamics of the systems are expanded in more than one direction giving rise to more complex attractors (Li et al. 2005). The characteristics of hyperchaos can be underlined as follows: first, the minimal dimension of the phase space that embeds the hyperchaotic attractor should be at least four, which requires at least four coupled first-order autonomous ordinary differential equations (Rossler 1979). Second, the number of terms in the coupled equations that give rise to instability should be at least two, in which one of them should be a non-linear function (Rossler 1979). Third, any hyperchaotic system should have two positive Lyapunov exponents along with one zero and one negative Lyapunov exponents (Wolf et al. 1985b; Rossler 1979).

Earlier studies have shown that it is possible to control chaos in JJ systems both theoretically and experimentally using various ways, for instance, giving a feed back (Wang et al. 2006; Singer et al. 1991), applying a weak periodic force (Balanov et al. 2005; Braiman and Goldhirsch 1991; Soong et al. 2004). In all of those methods, dynamic dependencies to the system parameters are vital since the desired behavior of the JJ can be determined according to this process. Thus one can operate the device either in regular region or in highly-nonlinear regimes. In this context, Nayak and Kuriakose justified a chaos-hyperchaos-chaos transition scheme in an RF-biased coupled junction (Nayak and Kuriakose 2007).

In this section, an analytical hyperchaos analysis of the four-dimensional continuous system is conducted. Initially, the basic form of the equations can be written as
(Kurt and Canturk 2010),

$$
\begin{aligned}
\dot{x} &= y \\
\dot{y} &= -g_1(y)y - \sin(x) - \gamma(y + u) + i_1, \\
\dot{z} &= u, \\
\dot{u} &= -g_2(u)u - \Gamma \sin(z) - \gamma\alpha_0(y + u) + \alpha i_2
\end{aligned}
\tag{5.18}
$$

the equilibrium point of the system

$$
S_1 = (\arcsin(i_1), 0, \arcsin(\alpha i_2/\Gamma), 0).
\tag{5.19}
$$

Notice that the new dependent variables (x, y, z, u) in Eq. (5.18) are the redefined
from the variables $(\phi_1, v_1, \phi_2, v_2)$ in Eqs. (5.12)–(5.15). By using this point, one can
easily define the Jacobian matrix at S_1 in the form of

$$
J = \begin{pmatrix}
0 & 1 & 0 & 0 \\
-\sqrt{1 - i_1^2} & -g_1(0) - \gamma & 0 & -\gamma \\
0 & 0 & 0 & 1 \\
0 & -\alpha_0\gamma & -\sqrt{\Gamma^2 - \alpha_0^2 i_2^2} & -g_2(0) - \alpha_0\gamma
\end{pmatrix}.
\tag{5.20}
$$

In order to describe the bifurcation scenario, one has to define a bifurcation parameter.
In our case, γ is considered as the main bifurcation parameter. Eigenvalues λ of matrix
can be found from the equation

$$
-\lambda P - Q = 0.
\tag{5.21}
$$

Here

$$
\begin{aligned}
P &= (-g_1 - \gamma - \lambda)\{\lambda(g_2 + \alpha_0\gamma + \lambda) + \sqrt{\Gamma^2 - \alpha_0^2 i_2^2}\} + \gamma^2\alpha_0\lambda, \\
Q &= -\sqrt{1 - i_1^2}\{\lambda(g_2 + \alpha_0\gamma + \lambda) + \sqrt{\Gamma^2 - \alpha_0^2 i_2^2}\}
\end{aligned}
\tag{5.22}
$$

is defined.

5.4.1 Bifurcation Analysis

In this section, we will explore the bifurcation at the equilibrium point $S_1(sin^{-1}(i_1),$
$0, sin^{-1}(\alpha_0 i_2/\Gamma), 0)$. In the following, we consider γ as a bifurcation parameter. With
this respect, one should take bifurcation parameter $\gamma = 0$ in Jacobian Eq. (5.21) in

order to find out the bifurcation for γ. Thus we arrive at (Kurt and Canturk 2010)

$$(\lambda(g_1 + \lambda) + \sqrt{1 - i_1{}^2})(\sqrt{\Gamma^2 - \alpha^2 i_2{}^2} + \lambda(g_2 + \lambda)) = 0 \qquad (5.23)$$

after a basic analytical manipulation over the Jacobian. From Eq. (5.23), the eigenvalues can be found as,

$$\lambda_1 = -\frac{g_1}{2} - \sqrt{\frac{g_1^2}{4} - \sqrt{1 - i_1{}^2}} = -\frac{g_1}{2} - \sqrt{\Delta_1},$$

$$\lambda_2 = -\frac{g_1}{2} + \sqrt{\frac{g_1^2}{4} - \sqrt{1 - i_1{}^2}} = -\frac{g_1}{2} + \sqrt{\Delta_1},$$

$$\lambda_3 = -\frac{g_2}{2} - \sqrt{\frac{g_2^2}{4} - \sqrt{\Gamma^2 - \alpha_0^2 i_2{}^2}} = -\frac{g_2}{2} - \sqrt{\Delta_2},$$

$$\lambda_4 = -\frac{g_2}{2} + \sqrt{\frac{g_2^2}{4} - \sqrt{\Gamma^2 - \alpha_0^2 i_2{}^2}} = -\frac{g_2}{2} + \sqrt{\Delta_2}. \qquad (5.24)$$

Then using the transmission matrix, the system of Eq. (5.18) becomes

$$\begin{pmatrix} \dot{x}_1 \\ \dot{x}_2 \\ \dot{x}_3 \\ \dot{x}_4 \end{pmatrix} = \begin{pmatrix} 0 & \frac{2 + \frac{g_1}{\sqrt{\Delta_1}}}{4\sqrt{\Delta_1} - 2g_1} & 0 & 0 \\ 0 & \gamma\left(-\frac{1}{2} + \frac{g_1}{4\sqrt{\Delta_1}}\right) & 0 & 0 \\ 0 & 0 & 0 & -\frac{(2\sqrt{\Delta_2} + g_2)^2}{8\sqrt{\Delta_2}(\Gamma^2 - \alpha_0^2 i_2{}^2)} \\ 0 & 0 & 0 & -\alpha_0\gamma\left(\frac{1}{2} - \frac{g_2}{4\sqrt{\Delta_2}}\right) \end{pmatrix} \begin{pmatrix} x_1 \\ x_2 \\ x_3 \\ x_4 \end{pmatrix} + \begin{pmatrix} k_0 \\ k_1 \\ k_2 \\ k_3 \end{pmatrix},$$

k_0, k_1, k_2 and k_3 are defined as,

$$k_0 = \left(\frac{1}{2} + \frac{g_1}{4\sqrt{\Delta_1}}\right)\left(i_1 - (x_1 + x_2)g_1[x_1 + x_2]\right. $$
$$+ \sin\left(\frac{x_1}{\sqrt{\Delta_1} + \frac{g_1}{2}} - \frac{x_2}{\sqrt{\Delta_1} - \frac{g_1}{2}}\right)\Big), \qquad (5.25)$$

$$k_1 = \left(\frac{1}{2} - \frac{g_1}{4\sqrt{\Delta_1}}\right)\left(i_1 - (x_1 + x_2)g_1[x_1 + x_2]\right.$$
$$+ \sin\left(\frac{x_1}{\sqrt{\Delta_1} + \frac{g_1}{2}} - \frac{x_2}{\sqrt{\Delta_1} - \frac{g_1}{2}}\right)\Big), \qquad (5.26)$$

$$k_2 = \left(\frac{1}{2} + \frac{g_2}{4\sqrt{\Delta_2}}\right)\left(\alpha i_2 - (x_3 + x_4)g_2[x_3 + x_4]\right.$$

$$- \Gamma \sin \left(\frac{x_3 (\sqrt{\Delta_2} - \frac{g_2}{2})}{\sqrt{\Gamma^2 - \alpha_0^2 i_2^2}} + \frac{x_4 (-\sqrt{\Delta_2} - \frac{g_2}{2})}{\sqrt{\Gamma^2 - \alpha_0^2 i_2^2}} \right) \right), \tag{5.27}$$

$$k_3 = \left(\frac{1}{2} - \frac{g_2}{4\sqrt{\Delta_2}} \right) \left(\alpha_0 i_2 - (x_3 + x_4) g_2 [x_3 + x_4] \right.$$

$$- \Gamma \sin \left(\frac{x_3 (\sqrt{\Delta_2} - \frac{g_2}{2})}{\sqrt{\Gamma^2 - \alpha_0^2 i_2^2}} + \frac{x_4 (-\sqrt{\Delta_2} - \frac{g_2}{2})}{\sqrt{\Gamma^2 - \alpha_0^2 i_2^2}} \right) \right). \tag{5.28}$$

From now on, we will use the new variables x_1, x_2, x_3 and x_4 which were obtained after the Jacobian transformation of (x, y, z, u). Since the quantities k include all system dimensions, one should examine the stability of the equilibrium point S_1 near $\gamma = 0$. From the central manifold theory, the stability of S_1 at the vicinity $\gamma = 0$ can be determined by studying a one-parameter family of first-order ordinary differential equations on a center manifold. If introduced

$$x_2 = h_1(x_1, \gamma) = a_1 x_1^2 + a_2 x_1 \gamma + a_3 \gamma^2 + h.o.t,$$
$$x_3 = h_2(x_1, \gamma) = b_1 x_1^2 + b_2 x_1 \gamma + b_3 \gamma^2 + h.o.t, \tag{5.29}$$
$$x_4 = h_3(x_1, \gamma) = c_1 x_1^2 + c_2 x_1 \gamma + c_3 \gamma^2 + h.o.t$$

in order to get the vector field on the center manifold. Recall from Kurt and Canturk (2010) that the center manifold must satisfy,

$$N(h(x, \gamma)) = Dh \cdot k_0 - Bh - k = 0, \tag{5.30}$$

where

$$h = \begin{pmatrix} h_1 \\ h_2 \\ h_3 \end{pmatrix}, \quad k = \begin{pmatrix} k_1 \\ k_2 \\ k_3 \end{pmatrix}, \quad B = \begin{pmatrix} \gamma \left(-\frac{1}{2} + \frac{g_1}{4\sqrt{\Delta_1}} \right) & 0 & 0 \\ 0 & 0 & -\frac{(2\sqrt{\Delta_2} + g_2)^2}{8\sqrt{\Delta_2 (\Gamma^2 - \alpha_0^2 i_2^2)}} \\ 0 & 0 & -\alpha_0 \gamma \left(\frac{1}{2} - \frac{g_2}{4\sqrt{\Delta_2}} \right) \end{pmatrix} \tag{5.31}$$

Finally, using above equations, we arrive at a vector field reduced to the center manifold,

$$\dot{x}_1 = \chi a_1 x_1^2 + \chi a_2 x_1 \gamma + \chi a_3 \gamma^2 + \xi \left(-a_1 g_1 x_1^2 + x_1 g_1 (1 + a_2 \gamma) \right.$$

$$+ a_3 g_1 \gamma^2 + \frac{x_1}{\sqrt{\Delta_1} + \frac{g_1}{2}} - \frac{a_1 x_1^2 + a_2 x_1 \gamma + a_3 \gamma^2}{\sqrt{\Delta_1} - \frac{g_1}{2}} \right), \tag{5.32}$$

$$\dot{\gamma} = 0. \tag{5.33}$$

Here, the terms χ and ξ are expressed by

$$\chi = \frac{2 + \frac{g_1}{\sqrt{\Delta_1}}}{4\sqrt{\Delta_1} - 2g_1}, \quad \xi = \frac{1}{2} + \frac{g_1}{4\sqrt{\Delta_1}}. \tag{5.34}$$

Note that we calculate the center manifold to the third order of accuracy. Simplifying Eq. (5.32), one arrives at the following equation,

$$\dot{x}_1 = Ax_1 + Bx_1{}^2. \tag{5.35}$$

This equation determines a transcritical bifurcation. However the stabilities strictly depend on the coefficients of A and B as follows:

$$A = \chi a_2 \gamma + \xi \left(\frac{1}{\sqrt{\Delta_1 + \frac{g_1}{2}}} - g_1 - a_2 g_1 \gamma - \frac{a_2 \gamma}{\sqrt{\Delta_1 - \frac{g_1}{2}}} \right), \tag{5.36}$$

$$B = \chi a_1 - \xi \left(a_1 g_1 + \frac{a_1}{\sqrt{\Delta_1 - \frac{g_1}{2}}} \right) \tag{5.37}$$

The stability of branches can be found by setting $x_1 \to x_1 + \delta$ into Eq. (5.32). Considering a characteristic exponent, the above-expression yields to,

$$\omega = (A - 2Bx_1). \tag{5.38}$$

In other words, $\omega = A$ for $x_1 = 0$ and $\omega = -A$ for $x_1 = A/B$. Let us now consider the following conditions for $(x_1, \gamma) = (\arcsin(i_1), 0)$. Equation (5.38) reads as

$$\omega = \frac{g_1 - g_1{}^3 + 2g_1\sqrt{1 - i_1{}^2} + \sqrt{g_1{}^2 - 4\sqrt{1 - i_1{}^2}} - g_1{}^2\sqrt{g_1{}^2 - 4\sqrt{1 - i_1{}^2}}}{g_1{}^2 - 4\sqrt{1 - i_1{}^2} + g_1\sqrt{g_1{}^2 - 4\sqrt{1 - i_1{}^2}}}$$
$$+ \frac{g_1 \arcsin(i_1)\left(\sqrt{g_1{}^2 - 4\sqrt{1 - i_1{}^2}} + g_1\left(2\sqrt{1 - i_1{}^2} - 1\right)\right)}{i_1\left(g_1{}^2 - 4\sqrt{1 - i_1{}^2} + g_1\sqrt{g_1{}^2 - 4\sqrt{1 - i_1{}^2}}\right)}. \tag{5.39}$$

For instance, using the above expression, we find out representative values for the eigenvalue at the point S_1. The sign of ω changes at $I_1 = 0.945951$ from the stable value of $\omega = -1.08537 \times 10^{-15} - 1.24798i$ to the unstable value of $1.58415 \times 10^{-15} - 1.24798i$. Note also that this solution also have a finite imaginary part driving the system to an oscillatory solution at this point. A detailed analysis also confirms the increasing trend of exponentials as will be stated in the following subsection.

5.4.2   Results of Hyperchaos Simulations

Since the JJ device shows a variety of dynamics, we restrict ourselves in this subsection only the hyperchaotic states. In order to identify the hyperchaotic behavior, one of the essential parameter, namely coupling resistance R_{cp} is taken into account. Note that the bifurcation parameter γ in the previous section is nothing else than the inverse of R_{cp}. Since it dominates the dynamics as a damping factor, Fig. 5.15 shows the variations of four Lyapunov exponents as function of R_{cp} (i.e. $V_o/(I_c\gamma)$). Since the present system given by Eq. (5.18) includes four first order equations and four corresponding Lyapunov exponents, Fig. 5.15 indicates the variations of these exponents (Kurt and Canturk 2010).

It is clear from the figure that the decrease in R_{cp} causes a high complexity in the system due to positive exponents. Strictly speaking, the maximal value of Lyapunov exponent increases up to $\lambda_m = 5.57$ at $R_{cp} = 350\Omega$ (i.e. $\gamma = 752$) which indicates a strong chaoticity inside the hyperchaotic region. For the moderate values of resistance (i.e. $R_{cp} \approx 660$ or $\gamma = 399$) a decrease to $\lambda_m = 0$ is the indicator of a regular behavior. Note that the present device exhibits a larger hyperchaotic region compared to the one in Nayak and Kuriakose (2007). Regime is transformed into a fixed-point attractor where negative exponents survive with increasing resistance (i.e. decreasing γ). Representative bifurcation diagrams are given in Fig. 5.16a, b in order to sketch out the variation of phases and voltages as function of the bifurcation parameter R_{cp}. Notice that phase differences ϕ_1, ϕ_2, voltages v_1, v_2 and the coupling current I_{cp} are given in the corresponding plots for the same parameter set in Fig. 5.15. Comparing Fig. 5.15 with Fig. 5.16a, b, we can state that the highest two exponents in the hyperchaotic region belongs to ϕ_1 and v_1, since the values of them rapidly go away from their central values at high R_{cp}s for $R_{cp} < 430\Omega$ (i.e. $\gamma = 612$) as seen in the corresponding diagrams. The diagrams in Fig. 5.16a, b also support the

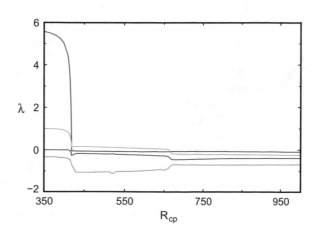

Fig. 5.15 Variation of Lyapunov exponents as function coupling resistance R_{cp}

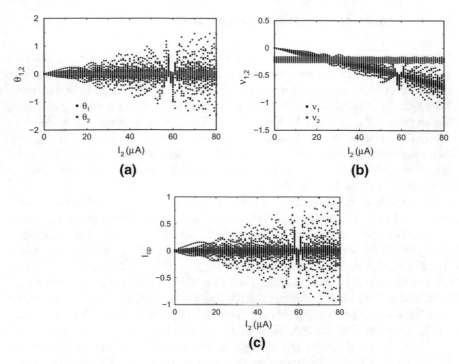

Fig. 5.16 Variation of phase and voltages as function of the parameters

results of Lyapunov exponents shown in Fig. 5.15, since ϕ_1 and v_1 are highly chaotic especially for lower R_{cp} values such as $R_{cp} = 430\Omega$. Since I_{cp} in Fig. 5.16b is formed by voltages v_1 and v_2, it reflects the characteristics of voltages. Therefore, the values of I_{cp} rapidly go away from their central values for lower R_{cp} values.

For further step, the effect of source current I_2 can also be investigated in detail. In principle, the contribution of I_1 on dynamics is similar to the contribution of I_2, hence we only focus on I_2. Figure 5.17 represents the bifurcation diagram for the parameter I_2. The complex dynamics becomes stronger due to the aperiodic and chaotic phases which are clearly seen for increasing I_2. While θ_2 which indicates the phase of the second junction stays around 0 rads, θ_1 shows a strong complexity in Fig. 5.17a. Similarly, the voltage values become more complex, when I_2 increases further as seen in Fig. 5.17b. Meanwhile, there exists a very narrow periodic window around $I_2 = 58 \mu A$ for θ_1 and v_1. In the case of I_{cp} shown in Fig. 5.17c, the dynamics is mostly dominated by v_1 similar to the case in Fig. 5.17b. The characteristic features of I_{cp} and v_1 resemble to each other. This similarity can also be seen in Fig. 5.16, since the trajectories on the v_1-I_{cp} phase plane intensify along the diagonal region due to the strong correlation between v_1 and I_{cp}. Figure 5.17c also states that the complexity increases drastically as function of I_2 similar to other bifurcation diagrams. In a recent study, Feng and Shen (2008) have also proven in a RCL-shunted JJ system that feeding current plays an important role in order to determine the

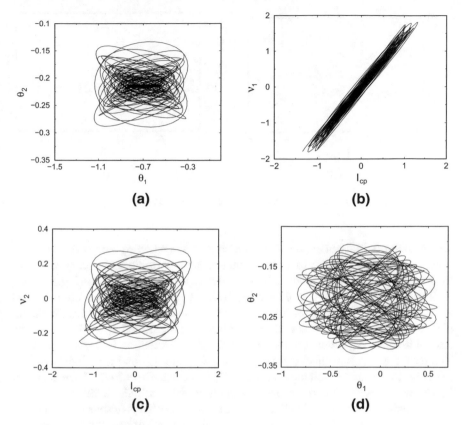

Fig. 5.17 Bifurcation diagram for the parameter I_2

dynamics. According to their simulations, a dramatical increase in λ values beyond a certain I value is obvious similar to findings presented in this section. However they did not investigate the existence of hyperchaos in their model system since the synchronization phenomena was only considered.

In Fig. 5.17, some attractors are shown for a specific value of $R_{cp} = 350\Omega$ inside the hyperchaotic region. It is clear that the phase space trajectories in Fig. 5.17a, b and d diverge faster for relatively lower R_{cp}. While the attractors of phases (i.e. $\phi_1 - \phi_2$ in Fig. 5.17a) and voltage-current (i.e. $I_{cp} - v_2$ in Fig. 5.17b) indicate similar behavior for the same parameter set, $I_{cp} - v_1$ attractor in Fig. 5.17c gives a lower complexity since the phase space trajectories are close to each other compared to other phase plane representations in Fig. 5.17. We also present another attractor for a slightly different parameter set in Fig. 5.17d in order to show the diversity of dynamics. Comparing this phase plane representation with the attractor in Fig. 5.17a, the role of source current on the dynamics can be clearly observed.

In order to explore the source current and junction capacitance dependences of Lyapunov exponents, Fig. 5.18 is produced. In this figure, Lyapunov spectra is plotted

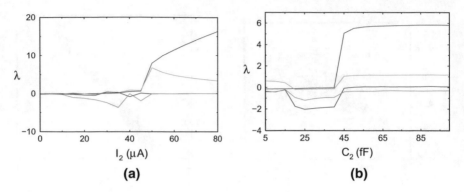

Fig. 5.18 Variation of Lyapunov exponents as function of I_2 and C_2

similar to Fig. 5.15. Four exponents characterize four system equations as mentioned before. According to Fig. 5.18a, the system is usually periodic for lower I_2 values for the selected parameter set. Thus, this figure also supports the findings in Fig. 5.16 due to decreasing complexity with decreasing I_2. The hyperchaotic regime arises around $I_2 = 27\,\mu A$ and survives for a wide range of source current. The reader should also notice that the dramatic rise in chaoticity occurs around $I_2 = 45\,\mu A$. In the case of junction capacitance shown in Fig. 5.18b, the situation is slightly different. The device exhibits a chaotic behavior for fairly low C_2 and a periodic window exists for moderate capacitances between $C_2 = 17$ and $C_2 = 38fF$. Then a sudden hyperchaotic case with two positive exponents reveals for high C_2 values.

The analysis shows (Kurt and Canturk 2010) that the effects of source currents can not be negligible in order to generate the desired dynamics of such a junction device. Note that the open regions on the surface of $I_1 - I_2$ indicate either fixed-point (i.e. $\lambda_m < 0$) or periodic (i.e. $\lambda_m = 0$) regions for the relevant parameter sets. As shown by calculations (Kurt and Canturk 2010), the device can be used in the field of secure communication and random number generation by adjusting wide ranges of system parameters. Strictly speaking, high feeding yield more complexity and randomness in the sense that the regular regions (i.e. open regions on the surface plot) are found for relatively lower current values forming two independent islands. Note that these two open regions correspond to either low I_1 or low I_2. Moreover sudden jumps to higher λ values are also obvious for moderate current values. These kinds of local hills occur as the natural results of index changes in maximal exponents. Such a local hill is also shown in Fig. 5.18a. To conclude, the effect of source currents for the birth of hyperchaos in such a system is indispensable, if the system is used for the secure communication or cryptological applications.

Fig. 5.19 Chaos
synchronization between
resistively coupled three JJ

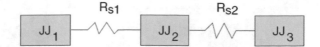

5.5   Chaos Synchronization Between JJs

As mentioned in previous sections, due to essential nonlinear dynamical properties
JJs study of chaos and its control in such systems could be useful from the application
point of view. Chaos control in such systems is important for JJ devices such as voltage
standards, detectors, SQUID, etc. where chaotic instabilities are not desired. Chaotic
JJs can be used for the short-distance secure communications, ranging purposes
(Dana et al. 2001). In a chaos-based secure communications, a message is masked
in the broadband chaotic output of the transmitter and synchronization between the
transmitter and receiver systems is used to recover a transmitted message (Scholl
and Schuster 2007; Boccaletti et al. 2002). In study Shahverdiev et al. (2014), it was
investigated chaos synchronization between JJs coupled unidirectionally with time
delay. It was demonstrated that the possibility of high-quality synchronization with
numerical simulations of such systems. For this purpose synchronization between
the following RC shunted JJs (Fig. 5.19) was considered and the system of equations
written in the dimensionless form has a form:

$$\dot{v}_1 + \frac{1}{\sqrt{\beta_{C1}}}v_1 + \sin\phi_1 = \{i_{dc1} + i_{01}\cos(\Omega_1 t + \theta_1)\}, \tag{5.40}$$

$$\dot{\phi}_1 = v_1$$

$$\dot{v}_2 + \frac{1}{\sqrt{\beta_{C2}}}v_2 + \sin\phi_2 = \{i_{dc2} + i_{02}\cos(\Omega_2 t + \theta_2)\} - \alpha_{s1}\{v_2(t) - v_1(t - \tau_2)\},$$

$$\dot{\phi}_2 = v_2 \tag{5.41}$$

$$\dot{v}_3 + \frac{1}{\sqrt{\beta_{C3}}}v_3 + \sin\phi_3 = \{i_{dc3} + i_{03}\cos(\Omega_3 t + \theta_2)\} - \alpha_{s2}\{v_3(t) - v_2(t - \tau_3)\},$$

$$\dot{\phi}_3 = v_3 \tag{5.42}$$

where ϕ_1, ϕ_2 and ϕ_3 are the phases of the junctions 1, 2, and 3, respectively; β_C
is called the McCumber parameter $(\beta_{C1,2,3}, R_{1,2,3})^2 = \hbar(2eI_{c1,c2,c3}C_{1,2,3})^{-1}$, where
$I_{c1,c2,c3}$, $R_{1,2,3}$ and $C_{1,2,3}$ are the junctions' critical current, the junction resistance,
and capacitance, respectively; $i_{dc1,dc2,dc3}$ is the driving the junctions DC; $i_{01,02,03}$
$\cos(\Omega_{1,2,3}t + \theta_{1,2,3})$ is the driving AC (or RF) current with amplitudes $i_{01,02,03}$, frequencies 1,2,3 and phases $\theta_{1,2,3}$; τ_1 and τ_2 are the coupling delay times between the
junctions 1–2 and 2–3, respectively; coupling between the junctions 1–2 and 2–3 is
due to the currents flowing through the coupling resistors R_{s_1} (between the junctions
1 and 2) and R_{s_2} (between the junctions 2 and 3); $\alpha_{s_1,s_2} = R_{1,2}/\sqrt{\beta_{C_1}\beta_{C_2}}R_{s_1,s_2}^{-1}$ are
the coupling strengths between junctions 1–2 and junctions 2–3. We note that in

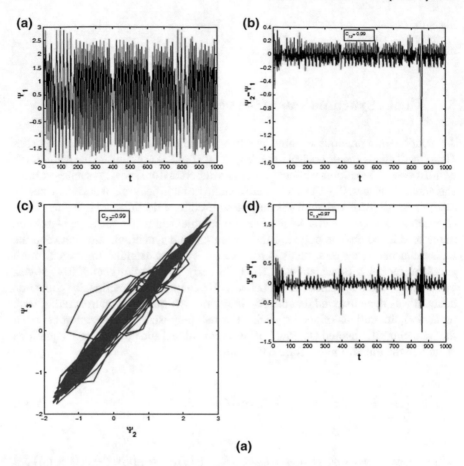

Fig. 5.20 Dynamic of variables: $\Psi_1(t)$ (a); $\Psi_2-\Psi_1$ (b); $\Psi_3(\Psi_2)$ (c); $\Psi_3-\Psi_1$

Eqs. (5.40)–(5.42), DC and AC current amplitudes are normalized with respect to the critical currents for the relative JJs; AC current frequencies 1, 2, 3 are normalized with respect to the JJ plasma frequency $\Omega^2_{J_1,J_2,J_3}(0) = 2eI_{c1,c2,c3}/(\hbar C_{1,2,3})$, and dimensionless time is normalized to the inverse plasma frequency. It is noted that if the external drive current is purely DC, then we have a second-order autonomous system, Eqs. (5.40)–(5.42) and therefore chaotic dynamic for Eqs. (5.40)–(5.42) is ruled out. In order to make the dynamics of the JJ Eqs. (5.40)–(5.42) chaotic, one can add AC current to the external driving, which makes Eqs. (5.40)–(5.42) a second-order nonautonomous system. Using numerical simulation in study (Shahverdiev et al. 2014) it was demonstrated that three unidirectionally time-delay coupled JJs can be synchronized. For the study of chaos synchronization between junctions using the correlation coefficient $C(\Delta t)$ (Haykin 1994)

Fig. 5.21 Dependence of correlation coefficients C_{12}, C_{23}, C_{13} on the delay time

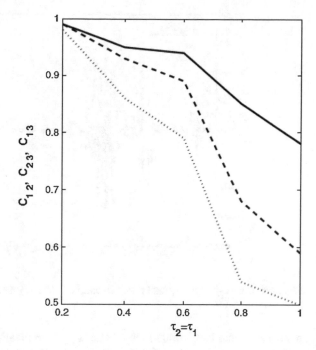

$$C(\Delta t) = \frac{\langle (x(t) - \langle x \rangle)\, (y(t + \Delta t) - \langle y \rangle) \rangle}{\sqrt{\langle (x(t) - \langle x \rangle)^2 \rangle \langle (y(t + \Delta t) - \langle y \rangle)^2 \rangle}}, \tag{5.43}$$

where x and y are the outputs of the interacting systems; the brackets $\langle . \rangle$ represent the time average; t is a time shift between the systems outputs: in the referred situation $t = 0$. This coefficient indicates the quality of synchronization: $C(\Delta t) = 1$ means perfect synchronization.

Figure 5.20a demonstrates dynamics of variable ψ_1; Fig. 5.20b pictures the error dynamics $\psi_2 - \psi_1$; C_{12} is the correlation coefficient between ψ_2 and ψ_1. Figure 5.20c represents the dependence between variables ψ_3 and ψ_2 with C_{23} being the correlation coefficient between ψ_2 and ψ_3. In Fig. 5.20d, it was presented the error dynamics between the end junctions $\psi_3 - \psi_1$, while C_{23} indicates the correlation coefficient between ψ_1 and ψ_3. The simulation results indicated in Shahverdiev et al. (2014) underlines the high quality synchronization between three identical JJs with different initial conditions.

In study Shahverdiev et al. (2014) the effect of the coupling delay time on the synchronization quality also was investigated. It is worth noting that synchronization quality is very sensitive to the value of the coupling delay. As demonstrated by the numerical simulations with increasing coupling delays between the JJs, correlation coefficients decays rapidly. In Fig. 5.21, it was presented the dependence of the correlation coefficients C_{12}, C_{23}, and C_{13} on the time delay. Note that one can increase the value of delay times τ_1 and τ_2 by keeping them identical. Figure 5.22 shows the

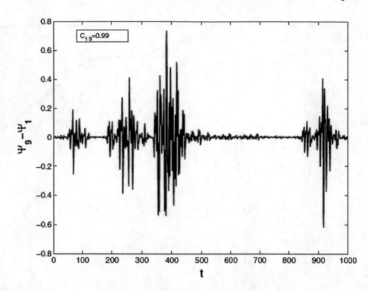

Fig. 5.22 Chaos synchronization between 9 coupled JJ Ψ_9–Ψ_1 and C_{19}

results of numerical simulations of chaos synchronization between 9 coupled JJs with $\tau_1 = \tau_2 = \cdots = \tau_8 = 0.001$, and the other parameters are as in Fig. 5.21: error dynamics of $\psi_9 - \psi_1$: C_{19} is the correlation coefficient between ψ_1 and ψ_9.

References

M. Abramovitz, A. Stegun, *Handbook on Mathematical Functions* (Dover, New York, 1972)

A. Abrikosov, *Fundamentals Theory of Metals* (North Holland, Amsterdam, 1988)

A. Abrikosov, L. Gorkov, I. Dzyaloshinski, *Methods of Quantum Field Theory in Statistical Physics* (Dover Book, Mineola, 1975)

V. Adler, C.H. Cheah, K. Gaj, D. Brock, E. Friedman, I.E.E.E. Trans, Appl. Supercond. **7**(2), 3294 (1997)

D.F. Agterberg, E. Demler, B. Janko, Phys. Rev. B **66**(21), 214507 (2002)

H. Akoh, S. Sakai, A. Yagi, H. Hayakawa, Jpn. J. Appl. Phys. 22 (Part 2, No. 7), L435 (1983)

B.P. Algul, I. Avci, R. Akram, A. Bozbey, M. Tepe, D. Abukay, in *Proceedings of the AIP Conference*, vol. 899 (AIP Publishing, Melville, 2007), pp. 762–762

V. Ambegaokar, A. Baratoff, Phys. Rev. Lett. **10**(11), 486 (1963)

M.H.S. Amin, A.N. Omelyanchouk, A. Zagoskin, Physica C **372–376**, 178 (2002)

M.H.S. Amin, A.Y. Smirnov, A.M. Zagoskin, T. Lindström, S.A. Charlebois, T. Claeson, A.Y. Tzalenchuk, Phys. Rev. B **71**(6), 064516 (2005)

W. Anacker, IBM, J. Res. Dev. **24**(2), 107 (1980)

A. Andreev, JETP **46**, 182 (1964)

I. Askerzade, Turk. J. Phys. **22**(8), 811 (1998)

I. Askerzade, *Unconventional Superconductors: Anisotropy and Multiband Effects* (Springer, Berlin, 2012)

I.N. Askerzade, Tech. Phys. **43**(9), 1123 (1998)

I.N. Askerzade, Tech. Phys. **45**(1), 66 (2000a)

I.N. Askerzade, Tech. Phys. **45**(7), 937 (2000b)

I.N. Askerzade, Tech. Phys. **48**(11), 1496 (2003)

I.N. Askerzade, Tech. Phys. Lett. **31**(8), 680 (2005a)

I.N. Askerzade, Tech. Phys. Lett. **31**(7), 622 (2005b)

I.N. Askerzade, Tech. Phys. **56**(5), 744 (2011)

I.N. Askerzade, S.E. Amrahov, Tech. Phys. Lett. **36**(2), 93 (2010)

I.N. Askerzade, V.K. Kornev, Radiotekhnika i Elektronika **39**, 869 (1994a)

I.N. Askerzade, V.K. Kornev, Radiotekhnika i Elektronika **39**, 510 (1994b)

I.N. Askerzade, T.V. Filipov, V.K. Kornev, Weak superconductivity (1990)

L. Aslamazov, A. Larkin, Sov. Phys. JETP Lett. **9**, 150 (1969)

P.K. Atanasova, T.L. Boyadjiev, Y.M. Shukrinov, E.V. Zemlyanaya. arXiv:1005.5691 [cond-mat] (2010a)

© Springer International Publishing AG 2017

I. Askerzade et al., *Modern Aspects of Josephson Dynamics and Superconductivity Electronics*, Mathematical Engineering, DOI 10.1007/978-3-319-48433-4

P.K. Atanasova, T.L. Boyadjiev, Y.M. Shukrinov, E.V. Zemlyanaya, P. Seidel, J. Phys, Conf. Series **248**(1), 012044 (2010b)

P.K. Atanasova, T.L. Boyadjiev, E.V. Zemlyanaya, Y.M. Shukrinov, in *Numerical Methods and Applications*, ed. by I. Dimov, S. Dimova, N. Kolkovska, 6046, (Springer, Berlin, 2010c), pp. 347–352

P. Bakhtin, P.E. Kandyba, V.I. Makhoa, G.Y. Pavlov. USSR Inventor's Cetrificate No 1008908 (1983)

A. Balanov, N. Janson, E. Scholl, E. Phys, E. Phys, Rev. E **71**, 016222 (2005)

A. Barone, G. Paterno, *Physics and Application of the Josephson Effect* (Wiley, New York, 1982)

T.H.K. Barron, W.T. Berg, J.A. Morrison, Proceedings of the Royal Society of London A: Mathematical. Phys. Eng. Sci. **250**(1260), 70 (1959)

T. Bauch, F. Lombardi, F. Tafuri, A. Barone, G. Rotoli, P. Delsing, T. Claeson, Phys. Rev. Letters **94**(8), 087003 (2005)

T. Bauch, T. Lindstrom, F. Tafuri, G. Rotoli, P. Delsing, T. Claeson, F. Lombardi, Science **311**(5757), 57 (2006)

J.G. Bednorz, K.A. Muller, Zeitschrift für Physik B Condensed Matter **64**(2), 189 (1986)

C.W.J. Beenakker, Phys. Rev. Lett. **67**(27), 3836 (1991)

F.S. Bergeret, A.F. Volkov, K.B. Efetov, Rev. Mod. Phys. **77**(4), 1321 (2005)

A. Bjorck, G. Golub, SIAM Review **19**(1), 5 (1977)

A.M. Black-Schaffer, J. Linder, Phys. Rev. B **82**(18), 184522 (2010)

G. Blatter, V.B. Geshkenbein, L.B. Ioffe, Phys. Rev. B **63**(17), 174511 (2001)

N. Bluzer, Phys. Rev. B **44**(18), 10222 (1991)

S. Boccaletti, J. Kurths, G. Osipov, D.L. Valladares, C.S. Zhou, Phys. Rep. **366**(1–2), 1 (2002)

N. Bogolyubov Jr., *A Method for Studying Model Hamiltonians* (Pergamon Press, New York, 1972)

N.N. Bogoliubov, Sov. Phys. JETP **34**(71), 41 (1958)

I.V. Borisenko, A.V. Shadrin, G.A. Ovsyannikov, I.M. Kotelyanskii, F.V. Komissinski, Tech. Phys. Lett. **31**(4), 332 (2005)

V. Bouchiat, D. Vion, P. Joyez, D. Esteve, C. Urbina, M.H. Devoret, Appl. Supercond. **6**(10–12), 491 (1999)

A. Bozbey, M. Fardmanesh, I.N. Askerzade, M. Banzet, J. Schubert, Supercond. Sci. Technol. **16**(12), 1554 (2003)

A. Bozbey, Y. Kita, K. Kamiya, M. Kozaka, M. Tanaka, T. Ishida, A. Fujimaki, IEICE Trans. Electron. **E99.C**(6), 676 (2016)

Y. Braiman, I. Goldhirsch, Phys. Rev. Lett. **66**(20), 2545 (1991)

E.N. Bratus, V.S. Shumeiko, E.V. Bezuglyi, G. Wendin, Phys. Rev. B **55**(18), 12666 (1997)

R.P. Brent, H.T. Kung, I.E.E.E. Trans, Comput. **31**(3), 260 (1982)

F.I. Buchholz, D. Balashov, M.I. Khabipov, D. Hagedorn, R. Dolata, R. Pöpel, J. Niemeyer, Phys. C **350**(3–4), 291 (2001)

D.A. Buck, Proc. IRE **44**(4), 482 (1956)

L.N. Bulaevskii, V.V. Kuzii, A.A. Sobyanin, JETP Lett. **25**, 290 (1977)

L.N. Bulaevskii, M.P. Maley, M. Tachiki, Phys. Rev. Lett. **74**(5), 801 (1995)

P. Bunyk, S. Rylov, in Proc. Ext. Abstracts 4th Int. Supercond. Electron. Conf. (ISEC93), Boulder, CO, 62 (1993)

P. Bunyk, D. Zinoviev, Appl. Supercond. IEEE Trans. **11**(1), 529 (2001)

P. Bunyk, K. Likharev, D. Zinoviev, Int. J. High Speed Electron. Syst. **11**(01), 257 (2001)

P. Bunyk, M. Leung, J. Spargo, M. Dorojevets, IEEE Trans. Appl. Supercond. **13**(2), 433 (2003). 00060

T.W. Button, N.M. Alford, Appl. Phys. Lett. **60**(11), 1378 (1992)

T.W. Button, P.A. Smith, G. Dolman, C. Meggs, S.K. Remillard, J.D. Hedge, S.J. Penn, N.M. Alford, I.E.E.E. Trans, Microw. Theory Tech. **44**(7), 1356 (1996)

A. Buzdin, J. Exp. Theor. Phys. Lett. **78**(9), 583 (2003)

A. Buzdin, M. Kupriyanov, JETP Lett. **53**, 321 (1991)

A.I. Buzdin, Rev. Mod. Phys. **77**(3), 935 (2005)

D. Cafagna, G. Grassi, Int. J. Bifurcation Chaos **13**(10), 2889 (2003)

N. Calander, T. Claeson, S. Rudner, Appl. Phys. Lett. **39**(6), 504 (1981)

A.O. Caldeira, A.J. Leggett, Phys. A **121**(3), 587 (1983a)

A.O. Caldeira, A.J. Leggett, Ann. Phys. **149**(2), 374 (1983b)

M. Canturk, I.N. Askerzade, I.E.E.E. Trans, Appl. Supercond. **21**(5), 3541 (2011)

M. Canturk, I.N. Askerzade, I.E.E.E. Trans, Appl. Supercond. **22**(6), 1400106 (2012)

M. Canturk, I.N. Askerzade, J. Supercond. Nov. Magn. **26**, 839 (2013)

M. Canturk, I. Askerzade, J. Supercond. Nov. Magn. **28**(2), 303 (2014)

M. Canturk, E. Kurt, I.N. Askerzade, COMPEL - Int. J. Comput. Math. Electr. Electron. Eng. **30**(2), 775 (2011)

D. Cassel et al., Phys. C **450**, 277 (2001)

A.B. Cawthorne, C.B. Whan, C.J. Lobb, J. Appl. Phys. **84**(2), 1126 (1998)

M. Celik, A. Bozbey, I.E.E.E. Trans, Appl. Supercond. **23**(3), 1701305 (2013)

M.E. Celik, A. Bozbey, J. Supercond. Nov. Magn. **26**(5), 1811 (2013)

H. Chaloupka, N. Klein, M. Peiniger, H. Piel, A. Pischke, G. Splitt, I.E.E.E. Trans, Microwave Theory Tech. **39**(9), 1513 (1991)

J.J. Chang, D.J. Scalapino, Phys. Rev. B **15**(5), 2651 (1977)

W. Chen, A.V. Rylyakov, V. Patel, J.E. Lukens, K.K. Likharev, I.E.E.E. Trans, Appl. Supercond. **9**(2), 3212 (1999)

B. Chesca, J. Low Temp. Phys. **110**(5–6), 963 (1998)

B. Chesca, Ann. Phys. (Leipzig) **8**(511) (1999)

B. Chesca, *in SQUIDs Handbook* (Wiley, Hoboken, 2004), p. 30

B. Chesca, R.R. Schulz, B. Goetz, C.W. Schneider, H. Hilgenkamp, J. Mannhart, Phys. Rev. Lett. **88**(17), 177003 (2002)

I. Chiorescu, P. Bertet, K. Semba, Y. Nakamura, C.J.P.M. Harmans, J.E. Mooij, Nature **431**(7005), 159 (2004)

D.C. Chung, I.E.E.E. Trans, Appl. Supercond. **11**(1), 107 (2001)

T. Clark, J. Baldwin, Electron. Lett. **3**(5), 178 (1967)

J. Clarke, in *SQUID Sensors: Fundamentals, Fabrication and Applications*, ed. by H. Weinstock. NATO ASI Series, vol. 329 (Springer, Netherlands, 1996), pp. 1–62

A.J. Dahm, A. Denenstein, T.F. Finnegan, D.N. Langenberg, D.J. Scalapino, Phys. Rev. Lett. **20**(16), 859 (1968)

S. Dana, D. Sengupta, K. Edoh, I.E.E.E. Trans, Circuits Syst. I Fundam. Theory Appl. **48**, 950 (2001)

V.V. Danilov, K.K. Likhaev, O.V. Snigirev, in *SQUID'80* ed. by H.-D. Hahlbohm, H. Lubbig (Berlin, 1980)

R. de Bruyn Ouboter, A.N. Omelyanchouk, E.D. Vol. Low Temp. Phys. **24**, 1017 (1998)

R. de Bruyn Ouboter, A.N. Omelyanchouk, Superlattice Microstruct. **23**, 1005 (1999)

E. Demler, A.J. Berlinsky, C. Kallin, G.B. Arnold, M.R. Beasley, Phys. Rev. Lett. **80**(13), 2917 (1998)

M.H. Devoret, J.M. Martinis, J. Clarke, Phys. Rev. Lett. **55**(18), 1908 (1985)

T. Dirks, T.L. Hughes, S. Lal, B. Uchoa, Y.F. Chen, C. Chialvo, P.M. Goldbart, N. Mason, Nat. Phys. **7**(5), 386 (2011)

Y.Y. Divin, I.M. Kotelyanski, P. Shadrin et al., Phys. C **282–287**, 115 (1997)

M. Dorojevets, P. Bunyk, D. Zinoviev, I.E.E.E. Trans, Appl. Supercond. **11**(1), 326 (2001)

M. Dorojevets, C.L. Ayala, N. Yoshikawa, A. Fujimaki, I.E.E.E. Trans, Appl. Supercond. **23**(3), 1700605 (2013a)

M. Dorojevets, C.L. Ayala, N. Yoshikawa, A. Fujimaki, I.E.E.E. Trans, Appl. Supercond. **23**(3), 1700104 (2013b)

R.F. Dziuba, B.F. Field, T.F. Finnegan, I.E.E.E. Trans, Instrum. Meas. **23**(4), 264 (1974)

R. Eberhart, J. Kennedy, in *Proceedings of the Sixth International Symposium on Micro Machine and Human Science MHS '95* (1995), pp. 39 –43

O. Edward, *Chaos in Dynamical Systems* (Cambridge University Press, Cambridge, 1993)

G. Eilenberger, Zeitschriftfur Physik **214**(2), 195 (1968)

D. Evans, S. Okolie, Comput. Math. Appl. **8**, 175 (1982)

R.L. Fagaly, Rev. Sci. Instrum. **77**(10), 101101 (2006)

E.S. Fang, T. Van Duzer, *Extension Abstract 2nd ISEC* (Tokyo, Japan, 1989), pp. 407–410

M. Fardmanesh, Appl. Opt. **40**(7), 1080 (2001)

M. Fardmanesh, I.N. Askerzade, Supercond. Sci. Technol. **16**(1), 28 (2003)

S.M. Faris, Appl. Phys. Lett. **36**(12), 1005 (1980)

Y.L. Feng, K. Shen, Eur. Phys. J. B **61**, 105 (2008)

A.K. Feofanov, V.A. Oboznov, V.V. Bol'ginov, J. Lisenfeld, S. Poletto, V.V. Ryazanov, A.N.
   Rossolenko, M. Khabipov, D. Balashov, A.B. Zorin, P.N. Dmitriev, V.P. Koshelets, A.V. Ustinov,
   Nat. Phys. **6**(8), 593 (2010)

L.V. Filippenko, V.K. Kaplunenko, M.I. Khabipov, V.P. Koshelets, K.K. Likharev, O.A. Mukhanov,
   S.V. Rylov, V.K. Semenov, A.N. Vystavkin, I.E.E.E. Trans, Magn. **27**(2), 2464 (1991)

T. Filippov, M. Dorojevets, A. Sahu, A. Kirichenko, C. Ayala, O. Mukhanov, I.E.E.E. Trans, Appl.
   Supercond. **21**, 847 (2011)

T.V. Filippov, A. Sahu, A.F. Kirichenko, I.V. Vernik, M. Dorojevets, C.L. Ayala, O.A. Mukhanov,
   Phys. Procedia **36**, 59 (2012)

C. Fourie, M. Volkmann, I.E.E.E. Trans, Appl. Supercond. **23**(3), 1300205 (2013)

C.J. Fourie, I.E.E.E. Trans, Appl. Supercond. **25**(1), 1 (2015)

J.R. Friedman, V. Patel, W. Chen, S.K. Tolpygo, J.E. Lukens, Nature **406**(6791), 43 (2000)

L. Fu, C.L. Kane, Phys. Rev. B **79**(16), 161408 (2009)

L. Fu, C.L. Kane, E.J. Mcle, Phys. Rev. Lett. **98**(10), 106803 (2007)

T. Fulton, J. Magerlein, L. Dunkleberger, I.E.E.E. Trans, Magn. **13**(1), 56 (1977)

Y.S. Galperin, A.T. Filippov, Sov. Phys. JETP **59**, 89 (1984)

I.A. Garifullin, D.A. Tikhonov, N.N. Garifyanov, L. Lazar, Y.V. Goryunov, S.Y. Khlebnikov, L.R.
   Tagirov, K. Westerholt, H. Zabel, Phys. Rev. B **66**(2), 020505 (2002)

V.B. Geshkenbein, A. Larlin, JETP Lett. **43**, 395 (1986)

V.B. Geshkenbein, A.I. Larkin, A. Barone, Phys. Rev. B **36**(1), 235 (1987)

V. Ginzburg, L. Landau, Sov. Phys. JETP **20**, 1064 (1950)

E. Goldobin, D. Koelle, R. Kleiner, A. Buzdin, Phys. Rev. B **76**(22), 224523 (2007)

A.A. Golubov, M.Y. Kupriyanov, Y.V. Fominov, JETP Lett. **75**, 190 (2002)

A.A. Golubov, M.Y. Kupriyanov, E. Ilichev, Rev. Mod. Phys. **76**(2), 411 (2004)

L.P. Gor'kov, V.Z. Kresin, Phys. C **367**(1–4), 103 (2002)

G. Grassi, S. Mascolo, J. Circuits Syst. Comput. **11**(01), 1 (2002)

Y.S. Greenberg, A. Izmalkov, M. Grajcar, E. Il'ichev, W. Krech, H.G. Meyer, M.H.S. Amin, A.M.
   van den Brink, Phys. Rev. B **66**(21), 214525 (2002)

N. Gronbech-Jensen, M.G. Castellano, F. Chiarello, M. Cirillo, C. Cosmelli, L.V. Filippenko, R.
   Russo, G. Torrioli, Phys. Rev. Lett. **93**(10), 107002 (2004)

R. Gross, A. Marx, *Applied Superconductivity* (Walther-Meissner-Institut-Garching, Munich, 2005)

V.N. Gubankov, K.I. Konstantinyan, V.P. Koshchelets, Sov. Phys. Tech. Phys. **27**, 1009 (1983)

A.L. Gudkov, V.K. Kornev, V.V. Makhov, Sov. Tech. Phys. Lett. **14**, 495 (1988)

A. Gurevich, M. Tachiki, Phys. Rev. Lett. **83**(1), 183 (1999)

C.A. Hamilton, F.L. Lloyd, R.L. Katz, I.E.E.E. Trans, Magn. **17**, 517 (1981)

C.A. Hamilton, F.L. Lloyd, K. Chieh, C. Goecke, I.E.E.E. Trans, Instrum. Meas. **38**, 314 (1989)

R.C. Hansen, I.E.E.E. Trans, Aerosp. Electron. Syst. **26**(2), 345 (1990)

P.K. Hansma, J. Appl. Phys. **44**(9), 4191 (1973)

D. Harris, in *Conference Record of the Thirty-Seventh Asilomar Conference on Signals, Systems
   and Computers 2004*, vol. 2 (2003), pp. 2213–2217

M.Z. Hasan, C.L. Kane, Rev. Mod. Phys. **82**(4), 3045 (2010)

S.G. Haupt, D.K. Lathrop, R. Matthews, S.L. Brown, R. Altman, W.J. Gallagher, F. Milliken, J.Z.
   Sun, R.H. Koch, I.E.E.E. Trans, Appl. Supercond. **7**(2), 2319 (1997)

H. Hayakawa, Jpn. J. Appl. Phys. **22**(S1), 437 (1983)

S. Haykin, *Communication Systems* (Wiley, New York, 1994)

A.M. Herr, M.J. Feldman, M.F. Bocko, Appl. Supercond. IEEE Trans. **9**(2), 3721 (1999)

M. Hidaka, T. Satoh, N. Ando, M. Kimishima, M. Takayama, S. Tahara, Physica C Part **2**(357–360), 1521 (2001)

M. Hidaka, S. Nagasawa, T. Satoh, K. Hinode, Y. Kitagawa, Supercond. Sci. Technol. **19**(3), S138 (2006)

D.S. Holmes, A.M. Kadin, M.W. Johnson, Computer **48**(12), 34 (2015)

J.Y. Hsieh, C.C. Hwang, A.P. Wang, W.J. Li, Int. J. Control **72**(10), 882 (1999)

B.A. Huberman, J.P. Crutchfield, N.H. Packard, Appl. Phys. Lett. **37**(8), 750 (1980)

Q. Hu, P.L. Richards, Appl. Phys. Lett. **55**(23), 2444 (1989)

T.L. Hwang, S.E. Schwarz, D.B. Rutledge, Appl. Phys. Lett. **34**(11), 773 (1979)

E. Ilichev, V. Zakosarenko, R.P.J. IJsselsteijn, H.E. Hoenig, H.G. Meyer, M.V. Fistul, P. Muller. Phys. Rev. B **59**(17), 11502 (1999)

E. Ilichev, M. Grajcar, R. Hlubina, Phys. Rev. Lett. **86**(23), 5369 (2001a)

E. Ilichev, V. Zakosarenko, L. Fritzsch, R. Stolz, H.E. Hoenig, H.G. Meyer, M. Gotz, A.B. Zorin, V.V. Khanin, A.B. Pavolotsky, J. Niemeyer, Rev. Sci. Instrum. **72**(3), 1882 (2001b)

E. Ilichev, H.E. Hoenig, H.G. Meyer, A.B. Zorin, V.V. Khanin, M. Gotz, A.B. Pavolotsky, J. Niemeyer, Phys. C **352**(1–4), 141 (2001c)

S. Intiso, I. Kataeva, E. Tolkacheva, H. Engseth, K. Platov, A. Kidiyarova-Shevchenko, I.E.E.E. Trans, Appl. Supercond. **15**(2), 328 (2005)

L.B. Ioffe, V.B. Geshkenbein, M.V. Feigel'man, A.L. Fauchere, G. Blatter, Nature **398**(6729), 679 (1999)

P.A. Ioselevich, M.V. Feigelman, Phys. Rev. Lett. **106**(7), 077003 (2011)

T. Ishida, N. Yoshioka, Y. Narukami, H. Shishido, S. Miyajima, A. Fujimaki, S. Miki, Z. Wang, M. Hidaka, J. Low Temp. Phys. **176**(3–4), 216 (2014)

ITRS, International Technology Roadmap for Semiconductors 2005 - Emerging Research Devices Technical Representative (2005)

A. Izmalkov, M. Grajcar, E. Il'ichev, T. Wagner, H.G. Meyer, A.Y. Smirnov, M.H.S. Amin, A.M. van den Brink, A.M. Zagoskin, Phys. Rev. Lett. **93**(3), 037003 (2004)

A. Izmalkov, S.H.W. van der Ploeg, S.N. Shevchenko, M. Grajcar, E. Il'ichev, U. Hübner, A.N. Omelyanchouk, H.G. Meyer, Phys. Rev. Lett. **101**(1), 017003 (2008)

Y. Izyumov, E. Kurmaev, *Superconductors Based on FeAs Compounds* (Springer, Berlin, 2010)

L.D. Jackel, J.P. Gordon, E.L. Hu, R.E. Howard, L.A. Fetter, D.M. Tennant, R.W. Epworth, J. Kurkijarvi, Phys. Rev. Lett. **47**(9), 697 (1981)

X.Y. Jin, J. Lisenfeld, Y. Koval, A. Lukashenko, A.V. Ustinov, P. Müller, Phys. Rev. Lett. **96**(17), 177003 (2006)

B.D. Josephson, Phys. Lett. **1**(7), 251 (1962)

JSIM and JSIM_n (1989)

K. Kadowaki, I. Kakeya, M.B. Gaifullin, T. Mochiku, S. Takahashi, T. Koyama, M. Tachiki, Phys. Rev. B **56**(9), 5617 (1997)

V. Kaplunenko, V. Borzenets, I.E.E.E. Trans, Appl. Supercond. **11**(1), 288 (2001a)

V. Kaplunenko, V. Borzenets, Appl. Supercond. IEEE Trans. **11**(1), 288 (2001b)

V.K. Kaplunenko, M.I. Khabipov, V.P. Koshelets, K.K. Likharev, O.A. Mukhanov, V.K. Semenov, I.L. Serpuchenko, A.N. Vystavkin, I.E.E.E. Trans, Magn. **25**(2), 861 (1989)

V.G. Karklinsh, EKh Khermanis, *Bilateral Transducers of Signals* (Zinatne, Riga, 1980)

S.F. Karmanenko, A.A. Semenov, I.A. Khrebtov, V.N. Leonov, T.H. Johansen, Y.M. Galperin, A.V. Bobyl, A.I. Dedoboretz, M.E. Gaevski, A.V. Lunev, R.A. Suris, Supercond. Sci. Technol. **13**(3), 273 (2000)

I. Kataeva, H. Akaike, A. Fujimaki, N. Yoshikawa, S. Nagasawa, N. Takagi, I.E.E.E. Trans, Appl. Supercond. **21**(3), 809 (2011)

R.L. Kautz, J. Appl. Phys. **52**(10), 6241 (1981)

J. Kennedy, R. Eberhart, in *Proceedings of IEEE International Conference on Neural Networks*, vol. 4 (1942)

M. Kennedy, R. Rovatti, G. Setti, in *Chaotic Electronics in Telecommunications*, ed. by Raton (CRC Press, Boca Raton, 2000)

N. Khare, *Handbook of High Temperature Superconducting Electronics* (Marsel Dekker Inc, NY, 2003)

V.A. Khlus, I.O. Kulik, Sov. Phys. - Tech. Phys. **20**(3), 283 (1975)

J.R. Kirtley, Rep. Prog. Phys. **73**(12), 126501 (2010)

T. Kitamura, M. Yokoyama, J. Appl. Phys. **69**(2), 821 (1991)

R. Kleiner, P. Muller, Phys. Rev. B **49**(2), 1327 (1994)

R. Kleiner, A.S. Katz, A.G. Sun, R. Summer, D.A. Gajewski, S.H. Han, S.I. Woods, E. Dantsker, B. Chen, K. Char, M.B. Maple, R.C. Dynes, J. Clarke, Phys. Rev. Lett. **76**(12), 2161 (1996)

R. Kleiner, D. Koelle, F. Ludwig, J. Clarke, Proc. IEEE **92**(10), 1534 (2004)

M. Klein, D.J. Herrel, IEEE J. Solid-State Circuits **13**, 593 (1978)

N.V. Klenov, V.K. Kornev, N.F. Pedersen, Physica C **435**(1–2), 114 (2006)

N.V. Klenov, N.G. Pugach, A.V. Sharafiev, S.V. Bakurskiy, V.K. Kornev, Phys. Solid State **52**(11), 2246 (2010)

S. Knowles, in *Proceedings on14th IEEE Symposium on Computer Arithmetic* (1999), pp. 30–34

D. Koelle, R. Kleiner, F. Ludwig, E. Dantsker, J. Clarke, Rev. Mod. Phys. **71**(3), 631 (1999)

P.M. Kogge, H.S. Stone, I.E.E.E. Trans, Comput. C- **22**(8), 786 (1973)

F.V. Komissinski et al., Low Temp. Phys. **20**, 599 (2004)

P. Komissinskiy, G.A. Ovsyannikov, I.V. Borisenko, Y.V. Kislinskii, K.Y. Constantinian, A.V. Zaitsev, D. Winkler, Phys. Rev. Lett. **99**(1), 017004 (2007)

V. Kornev, V. Semenov, I.E.E.E. Trans, Magn. **19**(3), 633 (1983)

V.K. Kornev, V.K. Semenov, Extended Abstracts of International Superconductivity Electronics Conference(ISEC'1987)(Tokyo, Japan, 1987) pp. 131–134 (1987)

V.K. Kornev, T.Y. Karminskaya, Y.V. Kislinskii, P.V. Komissinki, K.Y. Constantinian, G.A. Ovsyannikov, Phys. C **435**(1–2), 27 (2006)

T. Koyama, M. Tachiki, Phys. Rev. B **54**(22), 16183 (1996)

H.A. Kramers, Physica **7**(4), 284 (1940)

S.P. Kruchinin, V.F. Klepikov, V.E. Novikov, D.S. Kruchinin, Mater. Sci. Pol. **23**, 1009 (2005)

S.P. Kruchinin, V. Novikov, V. Klepikov, Metrol. Meas. Syst. **15**, 381 (2008)

I.O. Kulik, A.N. Omelyanchuk, Pis'ma Zh. Eksp. Teor. Fiz. 21, 216. JETP Lett. **21**, 96 (1975)

I.O. Kulik, A.N. Omelyanchuk, Sov. J. Low Temp. Phys. (Engl. Transl.); (United States) 4:3 (1978)

I.O. Kulik, I.K. Yanson, *Josephson Effect in Superconducting Tunneling Structures* (Wiley, New York, 1972)

J. Kunert, O. Brandel, S. Linzen, O. Wetzstein, H. Toepfer, T. Ortlepp, H.G. Meyer, I.E.E.E. Trans, Appl. Supercond. **23**(5), 1101707 (2013)

J. Kurkijarvi, J. Appl. Phys. **44**(8), 3729 (1973)

E. Kurt, M. Canturk, Physica D **238**(22), 2229 (2009)

E. Kurt, M. Canturk, Int. J. Bifurcation Chaos **20**(11), 3725 (2010)

L. Kuzmin, K. Likharev, E. Migulin, E. Polunin, N. Simonov, in *SQUIDs and Their Applications 1985*, ed. by H. Hahilbohm, H. Lubbig (Walter de Gruyter, Berlin, 1985), p. 1029

M.J. Lancaster, Z. Wu, T. Maclean, N.M. Alford, Cryogenics **30**, 1048 (1990)

M. Lancaster, Z. Wu, Y. Huang, T.S.M. Maclean, X. Zhou, C. Gough, N.M. Alford, Supercond. Sci. Technol. **5**(4), 277 (1992)

L.D. Landau, E.M. Lifshitz, *Quantum Mechanics. A Course of Theoretical Physics*, vol. 3 (Pergamon Press, Oxford, 1965)

V.N. Leonov, I.A. Khrebtov, Instr. Exp. Tech. **36**, 501 (1993)

Y. Li, W.K.S. Tang, G. Chen, Int. J. Circuit Theory Appl. **33**(4), 235 (2005)

K. Likharev, O. Mukhanov, V. Semenov, in *SQUID'85*, ed. by D. Hahlbohm, K. Lubbig (1985), pp. 1103–1108

K. Likharev, V. Semenov, A. Zorin, in Superconducting Devices, ed. by S.T. Ruggerio, D.A. Rudman, (Elsevier, Amsterdam, 1990), pp. 1–49

K.K. Likharev, Rev. Mod. Phys. **51**(1), 101 (1979)

K.K. Likharev, Sov. Phys. Uspekhi **26**(1), 87 (1983)

K. Likharev, *Dynamics of Josephson Junction and Circuits* (Gorden Breach Publ, New York, 1986)

K.K. Likharev, *In Encyclopedia of Materials: Science and Technology* (Elsevier, Amsterdam, 2001)

K.K. Likharev, Journal de Physique IV - Proceedings **12**(3), 1 (2002)

K.K. Likharev, Phys. C **482**, 6 (2012)

K.K. Likharev, V.K. Semenov, I.E.E.E. Trans, Appl. Supercond. **1**(1), 3 (1991)

K.K. Likharev, A.B. Zorin, J. Low Temp. Phys. **59**(3–4), 347 (1985)

S.Z. Lin, Phys. Rev. B **86**(1), 014510 (2012)

T. Lindstrom, J. Johansson, T. Bauch, F. Tafuri, G. Rotoli, P. Delsing, T. Claeson, F. Lombardi, Phys. Rev. B **74**, 1453 (2006)

S.V. Lotkhov, S.A. Bogoslovsky, A.B. Zorin, J. Niemeyer, Appl. Phys. Lett. **78**, 946 (2001)

W. Lum, H. Chan, T.V. Duzer, I.E.E.E. Trans, Magn. **13**(1), 48 (1977)

J.B. Majer, F.G. Paauw, A.C.J. ter Haar, C.J.P.M. Harmans, J.E. Mooij, Phys. Rev. Lett. **94**(9), 090501 (2005)

R.R. Mansour, S. Ye, V. Dokas, B. Jolley, W.C. Tang, C.M. Kudsia, I.E.E.E. Trans, Microw. Theory Tech. **48**(7), 1199 (2000)

J.M. Martinis, M.H. Devoret, J. Clarke, Phys. Rev. Lett **55**(15), 1543 (1985)

J.M. Martinis et al., Phys. Rev. Lett. **89**, 117901 (2002)

M. Maruyama, M. Hidaka, T. Satoh, Physica C 392–396. Part **2**, 1441 (2003)

D. Massarotti, L. Longobardi, L. Galletti, D. Stornaiuolo, D. Montemurro, G. Pepe, G. Rotoli, A. Barone, F. Tafuri, Low Temp. Phys. **38**(4), 263 (2012)

J. Matisoo, Appl. Phys. Lett. **9**(4), 167 (1966)

I.I. Mazin, D.J. Singh, M.D. Johannes, M.H. Du, Phys. Rev. Lett. **101**(5), 057003 (2008)

D.E. McCumber, J. Appl. Phys. **39**(7), 3113 (1968)

J. Mercereau, Rev. Phys. Appl. **5**(1), 13 (1970)

T.M. Mishonov, Phys. Rev. B **44**(21), 12033 (1991)

J. Mooij, T. Orlando, L. Levitov, L. Tian, C. van der Wall, S. Lloyd, Science **285**, 1036 (1999)

N. Mori, A. Akahori, T. Sato, N. Takeuchi, A. Fujimaki, H. Hayakawa, Physica C 357–360. Part **2**, 1557 (2001)

O.A. Mukhanov, I.E.E.E. Trans, Appl. Supercond. **21**(3), 760 (2011)

J. Nagamatsu, N. Nakagawa, T. Muranaka, Y. Zenitani, J. Akimitsu, Nature **410**(6824), 63 (2001)

S. Nagasawa, K. Hinode, T. Satoh, M. Hidaka, H. Akaike, A. Fujimaki, N. Yoshikawa, K. Takagi, N. Takagi, IEICE Trans. Electron. **E97-C**(3), 132 (2014)

K. Nakajima, H. Mizusawa, H. Sugahara, Y. Sawada, I.E.E.E. Trans, Appl. Supercond. **1**(1), 29 (1991)

Y. Nakamura, Y.A. Pashkin, J.S. Tsai, Nature **398**(6730), 786 (1999)

C.R. Nayak, V.C. Kuriakose, Phys. Lett. A **365**(4), 284 (2007)

M.A. Nielsen, I.L. Chuang, *Quantum Computation and Quantum Information* (Cambridge University Press, Cambridge, 2000)

J. Niemeyer, J.H. Hinken, E. Vollmer, L. Grimm, W. Meier, Metrologia **22**(3), 213 (1986)

Niobium Process - Hypres Inc. (2016)

A.O. Niskanen, J.P. Pekola, H. Seppa, Phys. Rev. Lett. **91**, 177003 (2003)

A. Omelyanchouk, E. Ilichev, S. Shevchenko, *Quantum Phenomena in Qubits* (Naukova Dumka, Kiev, 2013). (in Russian)

T.P. Orlando, J.E. Mooij, L. Tian, C.H. van der Wal, L.S. Levitov, S. Lloyd, J.J. Mazo, Phys. Rev. B **60**(22), 15398 (1999)

Y. Ota, M. Machida, T. Koyama, H. Matsumoto, Phys. Rev. Lett. **102**(23), 237003 (2009)

Y. Ota, M. Machida, T. Koyama, Phys. Rev. B **83**(6), 060503 (2011)

G. Ovsyannikov, K. Konstantinyan, Low Temp. Phys. **39**, 423 (2012)

M. Ozer, M. Eren, Y.Tukel Celik, A. Bozbey. Cryogenics **63**, 174 (2014)

T.S. Parker, L.O. Chua, Proc. IEEE **75**, 892 (1987)

D. Patil, O. Azizi, M. Horowitz, R. Ho, R. Ananthraman, in *18th IEEE Symposium on Computer Arithmetic (ARITH '07)* (2007), pp. 16–28

L. Pecora, Phys. World **9**(5), 17 (1996)

G. Perez, H.A. Cerdeira, Phys. Rev. Lett. **74**(11), 1970 (1995)

G.E. Peterson, R.P. Stawicki, N.M. Alford, Appl. Phys. Lett. **55**(17), 1798 (1989)

F. Piquemal, G. Geneves, Metrologia **37**(3), 207 (2000)

R. Poli, J. Kennedy, T. Blackwell, Swarm Intell. **1**(1), 33 (2007)

S.V. Polonsky, V.K. Semenov, P.N. Shevchenko, Supercond. Sci. Technol. **4**(11), 667 (1991)

S. Polonsky, P. Shevchenko, A. Kirichenko, D. Zinoviev, A. Rylyakov, Appl. Supercond. IEEE Trans. **7**(2), 2685 (1997)

U. Poppe, Y.Y. Divin, M.I. Faley, J.S. Wu, C.L. Jia, P. Shadrin, K. Urban, I.E.E.E. Trans, Appl. Supercond. **11**(1), 3768 (2001)

A. Porch, M.J. Lancaster, T.S.M. Maclean, C. Gough, N.M. Alford, I.E.E.E. Trans, Magn. **27**, 2948 (1991)

X.L. Qi, S.C. Zhang, Rev. Mod. Phys. **83**(4), 1057 (2011)

R.A. Richardson, S.D. Peacor, C. Uher, F. Nori, J. Appl. Phys. **72**(10), 4788 (1992)

R. Rifkin, B.S. Deaver, Phys. Rev. B **13**(9), 3894 (1976)

J. Robinson, Y. Rahmat-Samii, I.E.E.E. Trans, Antennas Propag. **52**(2), 397 (2004)

O.E. Rossler, Phys. Lett. A **71**(2), 155 (1979)

G. Rotoli, T. Bauch, T. Lindstrom, D. Stornaiuolo, F. Tafuri, F. Lombardi, Phys. Rev. B **75**(14), 144501 (2007)

J.M. Rowell, Phys. Rev. Lett. **11**(5), 200 (1963)

YuA Ryabinin, *Stroboscopic Oscillography* (Sov.Radio, Moscow, 1968)

T. Ryhanen, H. Seppa, R. Ilmoniemi, J. Knuutila, J. Low Temp. Phys. **76**(5–6), 287 (1989)

A.V. Rylyakov, K.K. Likharev, I.E.E.E. Trans, Appl. Supercond. **9**(2), 3539 (1999a)

A.V. Rylyakov, K.K. Likharev, Appl. Supercond. IEEE Trans. **9**(2), 3539 (1999b)

K. Saitoh, Y. Soutome, T. Fukazawa, K. Takagi, Supercond. Sci. Technol. **15**(2), 280 (2002)

E. Scholl, H.G. Schuster, *In Handbook of Chaos Control* (Wiley-VCH, Weinheim, 2007)

M. Schubert, G. Wende, T. May, L. Fritzsch, H.G. Meyer, Supercond. Sci. Technol. **15**(1), 116 (2002)

H. Schulze, R. Behr, J. Kohlmann, F. Muller, J. Niemeyer, Supercond. Sci. Technol. **13**(9), 1293 (2000)

B. Seeber, *Handbook of Applied Superconductivity*, vol. 2 (CRC Press, Boca Raton, 1998)

A.V. Sergeev, MYu. Reizer. Int. J. Mod. Phys. B **10**(06), 635 (1996)

E.M. Shahverdiev, L.H. Hashimova, P.A. Bayramov, R.A. Nuriev, J. Supercond. Nov. Magn. **27**, 2225 (2014)

S. Shapiro, Phys. Rev. Lett. **11**(2), 80 (1963)

M. Sigrist, T.M. Rice, J. Phys. Soc. Jpn. **61**(12), 4283 (1992)

A.H. Silver, J.E. Zimmerman, Phys. Rev. **157**(2), 317 (1967)

J. Singer, Y.Z. Wang, H.H. Bau, Phys. Rev. Lett. **66**(9), 1123 (1991)

O.V. Snigirev, Radiotekhnika i Elektronika,(Moscow) **29**(11), 2216 (1984)

A.H. Sonnenberg, G.J. Gerritsma, H. Rogalla, Phys. C **326–327**, 12 (1999)

C.Y. Soong, W.T. Huang, F.P. Lin, P.Y. Tzeng, Phys. Rev. E **70**(1), 016211 (2004)

J. Sprott, *Chaos and Times Series Analysis* (Oxford University Press, London, 2003)

T. Stojanovski, J. Pihl, L. Kocarev, I.E.E.E. Trans, Circuits Syst. I Fundam. Theory Appl. **48**(3), 382 (2001)

T. Sugiura, Y. Yamanashi, N. Yoshikawa, I.E.E.E. Trans, Appl. Supercond. **21**(3), 843 (2011)

A. Svidzinski, *Space-inhomogeneous Problems of Superconductivity* (Nauka, Moscow, 1982), p. 309 (in Russian)

E.T. Swartz, R.O. Pohl, Rev. Mod. Phys. **61**(3), 605 (1989)

M. Tachiki, T. Koyama, S. Takahashi, Phys. Rev. B **50**(10), 7065 (1994)

S.H. Talisa, M.A. Janocko, D.L. Meier, C. Moskowitz, R.L. Grassel, J. Talvacchio, P. LePage, D.C. Buck, R.S. Nye, S.J. Pieseski, G.R. Wagner, I.E.E.E. Trans, Appl. Supercond. **5**(2), 2291 (1995)

Y. Tanaka, Phys. Rev. Lett. **72**(24), 3871 (1994)

Y. Tanaka, S. Kashiwaya, Phys. Rev. B **56**(2), 892 (1997)

M. Tanaka, K. Takata, T. Kawaguchi, Y. Ando, N. Yoshikawa, R. Sato, A. Fujimaki, K. Takagi, N. Takagi, in *Proceedings of the 15th International Superconductive Electronics Conference (ISEC)* (2015), pp. 1–3

H.J.M. ter Brake, F.I. Buchholz, G. Burnell, T. Claeson, D. Crete, P. Febvre, G.J. Gerritsma, H. Hilgenkamp, R. Humphreys, Z. Ivanov, W. Jutzi, M.I. Khabipov, J. Mannhart, H.G. Meyer, J. Niemeyer, A. Ravex, H. Rogalla, M. Russo, J. Satchell, M. Siegel, H. Topfer, F.H. Uhlmann, J.C. Villegier, E. Wikborg, D. Winkler, A.B. Zorin, Phys. C **439**(1), 1 (2006)

M. Terabe, A. Sekiya, T. Yamada, A. Fujimaki, I.E.E.E. Trans, Appl. Supercond. **17**(2), 552 (2007)

M. Tinkham, *Introduction to Superconductivity* (Courier Corporation, North Chelmsford, 1996)

S.K. Tolpygo, Low Temp. Phys. **42**(5), 361 (2016)

S.K. Tolpygo, V. Bolkhovsky, T.J. Weir, A. Wynn, D.E. Oates, L.M. Johnson, M.A. Gouker, I.E.E.E. Trans, Appl. Supercond. **26**(3), 1 (2016)

C.C. Tsuei, J.R. Kirtley, Rev. Mod. Phys. **72**(4), 969 (2000)

Y. Tukel, A. Bozbey, C.A. Tunc, J. Supercond. Nov. Magn. **26**(5), 1837 (2012)

Y. Tukel, A. Bozbey, C. Tunc, I.E.E.E. Trans, Appl. Supercond. **23**(3), 1700805 (2013)

A. Uchida, P. Davis, S. Itaya, Appl. Phys. Lett. **83**(15), 3213 (2003)

V.S. Udaltsov, J.P. Goedgebuer, L. Larger, J.B. Cuenot, P. Levy, W.T. Rhodes, Opt. Spectrosc. **95**(1), 114 (2003)

K.D. Usadel, Phys. Rev. Lett. **25**(8), 507 (1970)

L.R. Vale, R.H. Ono, D.A. Rudman, I.E.E.E. Trans, Appl. Supercond. **7**(2), 3193 (1997)

K.A. Valiev, Physics-Uspekhi **48**, 1 (2008)

T. Van Duzer, C.W. Turner, **13**(07), 369 (1981)

D.J. Van Harlingen, Rev. Mod. Phys. **67**(2), 515 (1995)

B.J. Vleeming, F.J.C. van Bemmelen, M.R. Berends, A.N. Omelyanchouk, R. de Bruyn, Ouboter. Physica B **262**, 246 (1999)

R.F. Voss, R.A. Webb, Phys. Rev. Lett. **47**(4), 265 (1981)

T.J. Walls, T.V. Filippov, K.K. Likharev, Phys. Rev. Lett. **89**(21), 217004 (2002)

H.B. Wang, P.H. Wu, T. Yamashita, Appl. Phys. Lett. **78**(25), 4010 (2001)

G. Wang, X. Zhang, Y. Zheng, Y. Li, Physica A **371**(2), 260 (2006)

G. Wendin, V.S. Shumeiko, Low Temp. Phys. **33**(9), 724 (2007)

C.B. Whan, C.J. Lobb, M.G. Forrester, J. Appl. Phys. **77**(1), 382 (1995)

S. Whiteley, G. Hohenwarter, S. Faris, I.E.E.E. Trans, Magn. **23**(2), 899 (1987)

J.R. Williams, A.J. Bestwick, P. Gallagher, S.S. Hong, Y. Cui, A.S. Bleich, J.G. Analytis, I.R. Fisher, D. Goldhaber-Gordon, Phys. Rev. Lett. **109**(5), 056803 (2012)

P. Wolf, B.V. Zeghbroeck, U. Deutsch, I.E.E.E. Trans, Magn. **21**(2), 226 (1985a)

A. Wolf, J.B. Swift, H.L. Swinney, J.A. Vastano, Physica D **16**(3), 285 (1985b)

D. Xi-Jie, H. Yi-Fan, W. Yu-Ying, Z. Jun, W. Zhen-Zhu, Chin. Phys. Lett. **27**(4), 044401 (2010)

S. Xu, Y. Rahmat-Samii, I.E.E.E. Trans, Antennas Propag. **55**(3), 760 (2007)

T. Yamashita, S. Takahashi, S. Maekawa, Phys. Rev. B **73**(14), 144517 (2006a)

T. Yamashita, S. Takahashi, S. Maekawa, Appl. Phys. Lett. **88**(13), 132501 (2006b)

Y. Yamanashi, T. Nishigai, N. Yoshikawa, I.E.E.E. Trans, Appl. Supercond. **17**(2), 150 (2007)

Y.S. Yerin, A.N. Omelyanchouk, Low Temp. Phys. **36**(10), 969 (2010)

D. Yohannes, S. Sarwana, S.K. Tolpygo, A. Sahu, Y.A. Polyakov, V.K. Semenov, Appl. Supercond. IEEE Trans. **15**(2), 90 (2005)

S. Yorozu, Y. Kameda, H. Terai, A. Fujimaki, T. Yamada, S. Tahara, Physica C **378–381**, 1471 (2002)

B. Yu, J. Li, Fractals **09**(03), 365 (2001)

J. Zaiman, *Principles of the Solids* (Cambridge University Press, UK, 1995)

H.H. Zappe, Appl. Phys. Lett. **27**(8), 432 (1975)

J.E. Zimmerman, P. Thiene, J.T. Harding, J. Appl. Phys. **41**(4), 1572 (1970)

A.B. Zorin, Phys. Rev. Lett. **96**(16), 167001 (2006)

M.S. Zubayri, M.A. Sculli, *Quantum Optics* (Cambridge University Press, Cambridge, 2003)

A.A. Zubkov, M.Y. Kupriyanov, V.K. Semenov, Sov. J. Low Temp. Phys. **7**(11), 661 (1981)

Index

© Springer International Publishing AG 2017
I. Askerzade et al., *Modern Aspects of Josephson Dynamics
and Superconductivity Electronics*, Mathematical Engineering,
DOI 10.1007/978-3-319-48433-4

Printed in the United States
By Bookmasters